A. L. Timofeev

Digital holography

A. L. Timofeev

Digital holography

Holographic coding

ScienciaScripts

Imprint
Any brand names and product names mentioned in this book are subject to trademark, brand or patent protection and are trademarks or registered trademarks of their respective holders. The use of brand names, product names, common names, trade names, product descriptions etc. even without a particular marking in this work is in no way to be construed to mean that such names may be regarded as unrestricted in respect of trademark and brand protection legislation and could thus be used by anyone.

Cover image: www.ingimage.com

This book is a translation from the original published under ISBN 978-620-7-65193-1.

Publisher:
Sciencia Scripts
is a trademark of
Dodo Books Indian Ocean Ltd. and OmniScriptum S.R.L publishing group

120 High Road, East Finchley, London, N2 9ED, United Kingdom
Str. Armeneasca 28/1, office 1, Chisinau MD-2012, Republic of Moldova, Europe
Printed at: see last page
ISBN: 978-620-7-74710-8

CONTENT

Introduction

Nowadays, the application of mathematical models and methods of image processing increasingly goes beyond the field of image processing itself. First of all, this applies to digital holography. Holography, as a method of wavefront restoration, can be used not only for recording and restoring three-dimensional images of an object. The fundamental property of holography - hologram divisibility [1] (possibility to recover the full image of an object from a hologram fragment) - is of interest for noise-resistant coding of arbitrary messages. The property of divisibility can be effectively used when transmitting information over a communication channel with a high noise level and/or with insufficient signal level, when large fragments of the message can be distorted or lost.

In this connection, it is of interest to transfer the principles of holographic image processing to the coding of arbitrary digital data and to develop holographic methods of noise-tolerant coding that allow correcting multiple errors.

This is relevant also for Radio-over-fiber (RoF) systems using multi-position X-QAM modulation techniques in fiber channels, which are subject to high requirements in terms of the necessary high signal-to-noise ratio, low dispersion and other characteristics that can lead to significant signal distortion and, consequently, to distortion of the original constellation. Thus, the aforementioned property of holography will significantly improve the noise immunity of signals in both fiber and radio segments of RoF systems.

High noise immunity of the holographic method of information processing and the ability to detect a signal many times lower than the interference level have found application in radio holography in the construction of radar systems [2-4]. Ways to reduce the radio channel load during information transmission for the subsequent synthesis of hologram 3D-images without their unacceptable distortions are considered in [5].

To use holographic coding for error correction was proposed in [6]. The fundamental difference of this task from image processing tasks possessing internal redundancy and allowing acceptable loss of accuracy is the requirement of exact correspondence of the decoded data block to the initial one. The considered method is based on representation of the initial digital block of arbitrary data as an image and calculation of the interference pattern of the wavefront created by this image.

Information encoding (hologram modeling) and decoding (digital array restoration) requires rather large computational resources. The complexity of calculations can be significantly reduced if we use a unit position code rather than a binary one to represent the initial block of digital information. In this case, the optical object for which the hologram is constructed is a point source on a black background, and the information is embedded in the coordinates of a point on the object field. The result of coding is the simplest hologram - a Fresnel zone plate, the coordinates of the center of which carry the encoded information.

In [7] the possibility of using one-dimensional radio holograms in order to reduce time and material costs during transmission over a radio channel was proved. Modeling of holographic coding with holograms of different shapes and sizes has shown that the efficiency of the code using a matrix hologram (m, m) corresponds to the efficiency of the code with a linear one-dimensional hologram with the number of points m^2 . Therefore, to reduce computational complexity, it is rational to use the representation of input data as a one-dimensional array and form a linear one-dimensional hologram.

1 Holographic coding

1.1 Coding algorithm

The holographic coding method is based on mathematical modeling of a one-dimensional hologram created in virtual space by a wave from an object representing the input data block [8].

Linear one-dimensional holographic coding of arbitrary digital information differs from optical holography by the following factors:

- object is one-dimensional

- the object is not bound to spatial measurements, the unit of measurement of the object and hologram size is the wavelength of radiation

- the wave propagates without attenuation and is coherent at any distance.

Preliminary input block of digital data X, representing a k-bit binary code, is transformed into a secondary block A - a spatial one-dimensional object consisting of $n = 2^k$ points A(i), i = 1,..., n, the value of one of which is equal to 1, the rest are zeros: A(i) = 1 at i=X, A(i)=0 at i≠X. This transformation lays down information redundancy with the number of digits r = n - k and sets the code rate R = k / n. The secondary block has (n - 1) zeros and one one at the position given by the primary block. Thus, the input data block is used as the address of the unit position in the sequence of zeros of the unit position code of the secondary block. In optical interpretation, the object (secondary block) represents a black line with a single luminous dot whose position is determined by the primary block being encoded.

The formation of the hologram is done by constructing a Fresnel zone ruler.

The distance between the points A(i) is equal to d. The cell A(i) = 1 is the source of a spherical wave propagating in the analysis plane and characterized by the wavelength λ = d.

Let us consider the values of the spherical wave front in the plane of the object location on a line parallel to the object at a distance L, in n points with a step d (Fig. 1.1). The wave from the source propagates without attenuation and hits all elements of the one-dimensional array H(j), j = 1,..., n. The values of the wave front at the considered points of H(j) are determined by the phase of the incoming wave φ. The value of the wave from the element A(i) at the location point of the hologram element H(j) is equal to

$$H(i) = A(i)\sin\varphi(i,j)$$

(1)

where φ (i,j) is the wave phase of the element A(i) at the point H(j). The distance l(i,j) between points A(i) and H(j) is equal to

$$l(i,j) = \sqrt{L^2 + d^2(i-j)^2}$$

then φ (i,j) is the fractional part of the ratio l(i,j) to the wavelength:

$$\varphi(i,j) = \{l(i,j)/\lambda\}$$

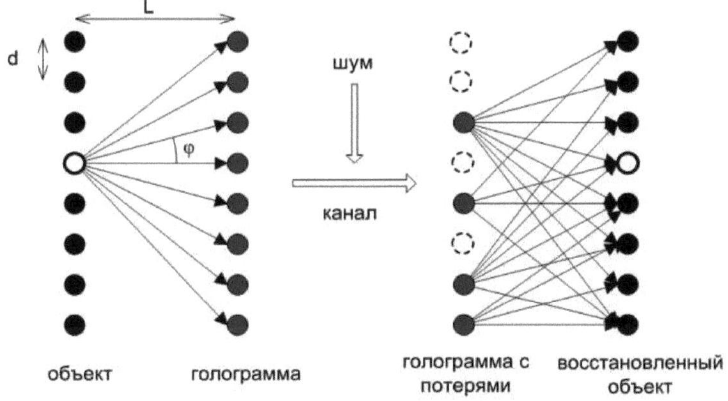

Figure 1.1. Spatial coding scheme

The set of n points H(j) forms the simplest linear hologram - a hologram of a point. Thus, a one-dimensional object A(i) corresponds to a one-dimensional hologram H(j). The values of the received hologram are rounded up to one bit - positive ones are taken as 1, negative ones - as 0, as a result of which an n-bit one-dimensional array H_O (j) is formed, which is a code combination corresponding to the k-bit input data block X. Rounding of the hologram values does not introduce digital rounding noise, since the information is embedded not in the point values (brightness) of the hologram H(j), but in the position (position number i=X) of the center of the Fresnel zones.

Thus, the array H_O (j) is a point hologram, aka a one-dimensional zone ruler, carrying information about the input data block in the form of an n-bit code of the Fresnel zone center coordinate.

The code combination (codeword) H_O (j) is transmitted over the communication channel with interference and decoding is performed on the receiving side - finding the center of Fresnel zones, determining its address and forming a k-bit output data block.

1.2 Decoding algorithm

The distorted code combination H_R (j) received over the communication channel is considered as a one-dimensional array of points, the number of which may be less than n due to the loss of some information, and the values of the received elements are distorted by noise. Each of the n points of the code combination H_R (j) is the source of a spherical wave with the same wavelength λ as in the encoding. The recoverable object A_R (i) is a one-dimensional array of points spaced at a step d on a straight line parallel to and at a distance L from the array H_R (j). If each of the n points of the array H_R (j) is represented by one bit, the object A_R (i) contains n multi-bit elements, the number of bits in which depends on the length of the code combination and the introduced redundancy r = n - k.

The intensity of the wave from the array point H_R (j) in the point of the reconstructed object A_R (i) is calculated in the same way as in coding (1). To each point of the object A_R (i) come waves from each point H_R (j) with its phase value and as a result of interference of these waves the values of the reconstructed object A_R (i) are formed:

$$A_R(i) = \sum_{i=1}^{n} H_R(j)\sin\varphi(i,j) , \quad \text{i=1,...,n.} \qquad (2)$$

Thus, the reconstructed object A_R (i) is a second-order hologram (hologram hologram) of the original object. From the optical point of view, the reconstructed object is a dark line with one bright point and a small background illumination in the other points.

To obtain the output data block in the form of a k-bit code, it is necessary to determine the address of the bright point - the number of position Y, where the maximum of the array A_R (i) is located. This number is the value of the output data block X_R. At signal distortion in the process of transmission (noise, interference, loss of a part of signal) the ratio of peak amplitude to noise level in the reconstructed signal decreases, and the result of decoding takes probabilistic character. Thus presence of any error in the decoded signal makes result of restoration unacceptable, therefore the basic characteristic of a code is probability of reception of error-free result.

1.3   Testing coding on a simulation model

Estimation of efficiency of holographic coding in comparison with existing noise-immune codes is carried out by modeling in MATLAB environment the process of distortion and restoration of code combination H_O (j) in the presence of random and packet errors.

Modeling Algorithm:

Coding:

- the size of the encoded block k

- the coded value X in the range 1,..., n is set, where n= 2^k

- a secondary block A(i), i=1,..., n, containing one unit in position i=X, the rest are zeros, is formed

- hologram H(j) is calculated by formula (1) and rounded to one bit, as a result a one-bit sequence H_O (j) is formed, which is a codeword for the initial value X.

Modeling of transmission over a noisy channel: using the awgn function built into MATLAB, white noise with a given signal-to-noise ratio (S/N) is superimposed on the signal H_O (j).

Decoding:

- The recovered object A_R (i) is calculated by formula (2)
- The maximum of A_R (i) is found, its coordinate i=Y is the output value.

Fig. 1.2 shows the result of encoding (one-dimensional hologram) of an 8-bit input data block with the value X=100, the length of the code combination is 256 bits.

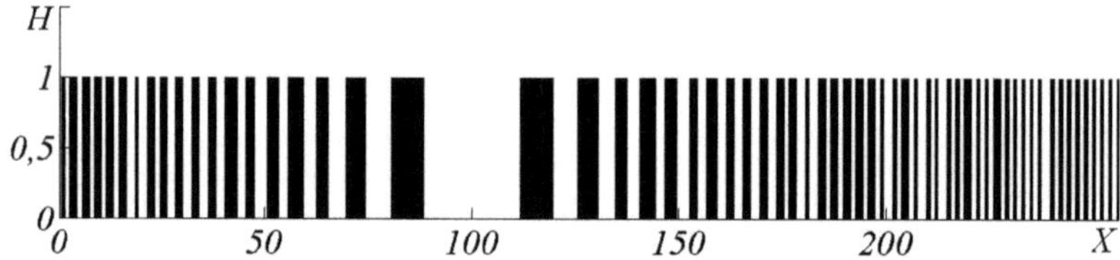

Fig. 1.2. One-dimensional hologram (sequence of zeros and ones) in a codeword of diynomial n=256

Fig. 1.3 shows the result of decoding (reconstructed image of the object) AR(i) with a pronounced maximum at position Y=100. Decoding was performed in the presence of noise in the channel, signal-to-noise ratio S/N=0 dB. Noise in the channel increases the parasitic illumination of the reconstructed object, while error correction occurs until the noise emission exceeding the information peak appears and, accordingly, the coordinate of the maximum Y changes.

The study of the dependence of the corrective ability of the holographic code on the distance between the object and the hologram L, the hologram pitch d and the wavelength λ has shown that the best results are achieved at the values L=nd and λ=d.

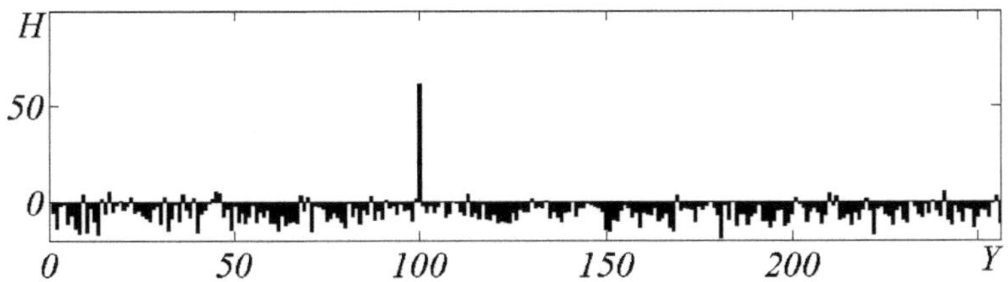

Figure 1.3. Decoding result at n=256, noise level S/N=0 dB, decoded value Y=100

Let's consider the error correction capabilities of the code.

Corrective abilities of a noise-resistant code having $n=2^k$ symbols in the code combination, of which k symbols are informational, are estimated by the maximum number of errors t, which it can correct at a given degree of redundancy. For example, for the Reed-Maller code (RM-code) $t=2^{k-2}$ -1 [9], which corresponds to the correction of errors of any kind, making up to 25% of the codeword length.

One of the most effective known codes for error correction is the Reed-Solomon code (RS-code), which is widely used in noise-resistant coding, in CD data recovery systems, and in creating archives with the possibility of restoring information in case of damage [10]. The limit of corrective ability (n,k) of PC-code is defined by Singleton's boundary [11], according to which to correct t errors the code should have at least n-k=2t check symbols, i.e. two check symbols per one error. At a large degree of redundancy (n>>k) the number of corrected errors t approaches 50% of the codeword length n. For example, PC-code (255,8) with a redundancy factor of 32, eliminates 123 errors, while the codeword contains 132 correct symbols - errors occupy 48% of the codeword. The peculiarity of the PC-code is that it demonstrates such a high correcting ability only for packet errors [10]. At the same time, the majority of digital channels described by the model of a binary symmetric channel without memory are characterized by random errors [10]. If we switch from packet errors to random errors uniformly distributed over the codeword, the maximum number of errors corrected by the PC-code will be t=n/2^k -1. It follows that the same variant of (n,k) PC-code at n=256, k=8 corrects 15 random errors, which is 6% of the codeword length.

The holographic code provides high resistance to both random and packet errors. To introduce a given number of random errors into the code combination, the randperm (m,m) function was used, generating m unique numbers in the range from 1 to m. These numbers were used as addresses of the bits where the error is introduced. The result of recovering a garbled code combination of length n=256 containing 80 random errors (31% of the codeword length was garbled) is shown in Figure 1.4. As a result of 100 000 tests it is established that probability of decoding error, i.e. probability of incorrect determination of information maximum position at 80 errors is 0,00005.

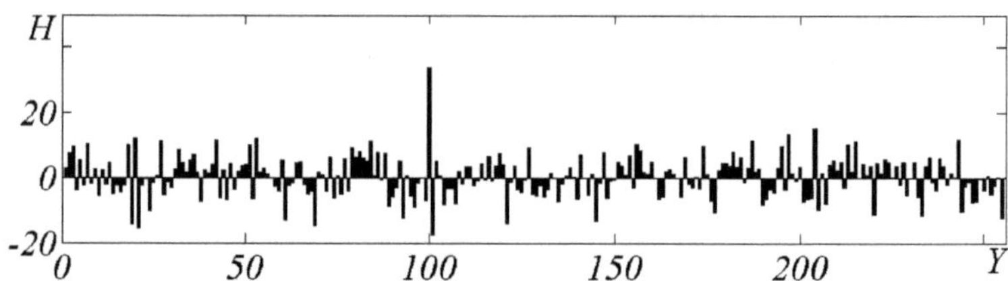

Figure 1.4. Decoding result of the codeword with 31% of random errors

For an estimation of noise immunity of a holographic code the reliability of information transmission on a channel with additive white Gaussian noise at use of several codes is compared. Dependence of decoding error probability on signal/noise ratio in a channel for PC-code, PM-code, majoritarian code and holographic code is considered. For this purpose the considered above marginal numbers of corrected errors for each code are taken and dependence of probability of occurrence of number of errors not more than marginal on a signal/noise ratio is constructed. In all cases the number of bits of the source word is 8, the length of the codeword is 256 bits (code rate R=1/32). The results are shown in Fig. 1.5.

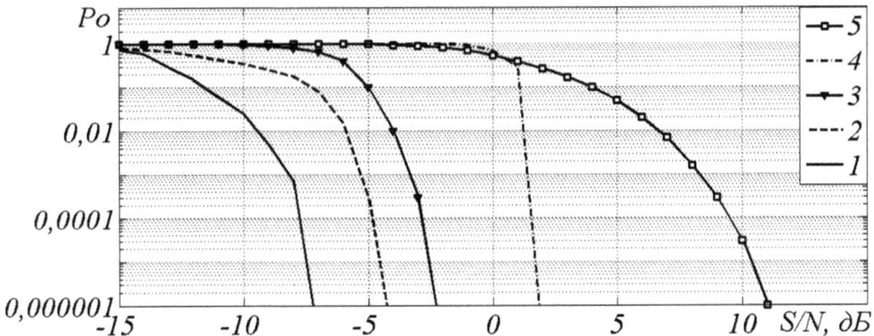

Fig. 1.5. Dependence of decoding error probability Po on signal/noise ratio at code rate R=1/32:

1 - holographic code, 2 - majority code, 3 - PM code,

4 - PC code, 5 - without coding

One of the most reliable ways of information transfer in strongly noisy channels is averaging within the introduced redundancy with the majoritarian way of a decision selection. However it has appeared that the holographic code is more noise-immune and provides gain on 2 dB in comparison with the majoritarian code that allows to receive probability of decoding error 10^{-6} at a signal/noise ratio S/N= -7 dB.

Additive noise exists in analog communication channels, and noise tolerant coding is used for digital signals in which the main channel quality characteristic is the bit error rate (BER).

Fig. 1.6 shows dependences of probability of decoding error P_O from probability of error in channel BER for the same codes, received by calculation for PC- and RM-codes, proceeding from a limit quantity of corrected errors for each code, and by modeling for majority and holographic codes. At the same code rate R=1/32 the holographic code provides probability of decoding error less than 10^{-4} at probability of error in the channel up to BER=0,34.

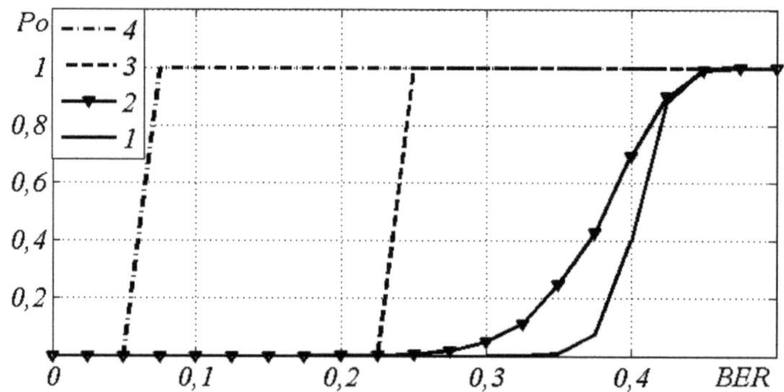

Fig. 1.6. Dependence of decoding error probability PO on error probability per bit BER: 1 - holographic code, 2 - majoritarian code, 3 - RM-code, 4 - PC-code

The considered examples of distortion of the transmitted message are characteristic for the binary symmetric channel without memory - errors in it are independent. At the same time, most radio communication channels (except for space channels) are characterized by the presence of fading effect due to multipath propagation. These include mobile wireless channels, ionospheric and tropospheric channels. In these channels distortions are caused by errors having the form of packets rather than individual isolated errors [10].

In itself the task of error correction in all bits of a binary codeword is trivial - for this purpose it is enough to invert each bit. The problem in this case for known codes is a choice from two equally probable results of decoding - direct and inverted. The holographic code unlike other codes gives the coinciding result of decoding both for the direct and inverted data block. Fig. 1.7 shows the result of decoding of the block with number of errors equal to 100%. As can be seen from comparison with the result of decoding of the undistorted block (Fig. 1.3), packet errors lead to inversion of a secondary hologram and for restoration of value of the input data block it is necessary only to define a point of a global extremum irrespective of its sign. This effect can also be explained by the fact that the coordinates of the center of the Fresnel zones are calculated with equal success for positive and negative (100% errors) Fresnel plate images.

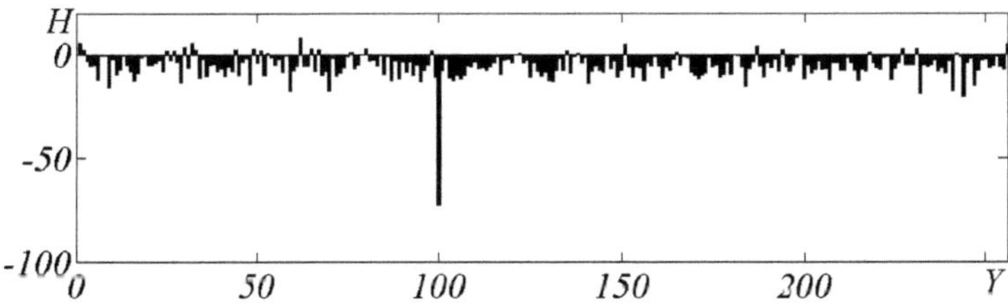

Fig. 1.7. Decoding result at 100% packet errors

It is more challenging to recover a block of data when the codeword contains about 50% of the bundled errors.

This problem is solved as follows. The decoded code combination is divided into four equal parts and each part is processed in direct and inverted form. In optical interpretation - each quarter of the hologram is considered as a positive and negative image, by each of which the center of Fresnel zones is determined - restoration of the initial object (value of the decoded data block). It allows to correct all packet errors with packet size from 0 to 100% of the codeword length at any packet location in the codeword. Thus, limiting capabilities of PC-code - correction of 50% of packet errors, of holographic code - 100% of errors.

The holographic code is also effective at decoding of signals containing a mixture of random and dependent errors. At research of decoder work in this mode errors in a code combination were introduced by means of the generator of random numbers. At number of errors less than 50% they are random and at codeword length n=1024 probability of decoding error P_O =0,001 at t=41% of errors (fig. 1.8).

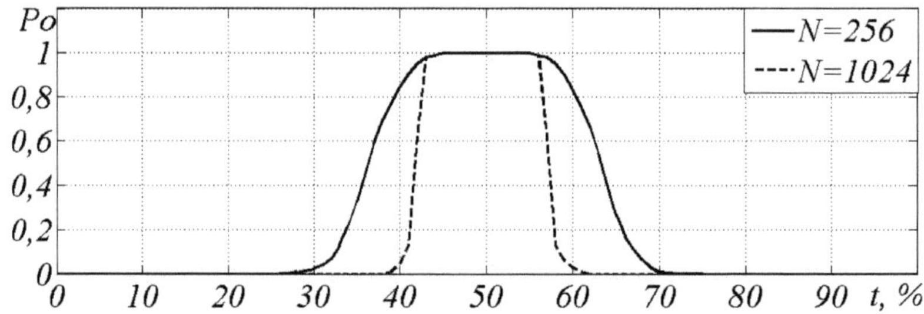

Fig. 1.8. Dependence of decoding error probability PO on number of errors t at code lengths n=256 and n=1024

The maximum possible number of independent random errors in a sufficiently long codeword combination is 50% of the number of bits in the codeword. When the source of random errors is further affected, the errors overlap and eventually a random codeword is formed, half of the bits of which coincide with the bits of any other codeword. For introduction of more than 50 % of errors bits for distortion are chosen taking into account already existing errors, therefore errors become dependent. At number of errors more than 50% decoding results form the right part of the graph of dependence of probability of decoding error P_O on number of errors t (fig. 1.8). At the total number of errors more than

60% the added errors fill the gaps between random errors, turning them into packet errors, that gives an opportunity to carry out the full recovery of information. Thus, the holographic code allows to correct a mixture of random and dependent errors if their total number exceeds 60% of the codeword length.

The holographic noise-tolerant coding method provides full recovery of data containing up to 40% random independent errors and up to 100% dependent packet errors.

Another advantage of the holographic code is less complexity of coding and decoding at variation of redundancy in wide limits. Decoding of the widely used Reed-Solomon code is a rather complicated problem, for the solution of which several types of algorithms have been developed. For example, the Peterson-Horenstein-Zirler algorithm reduces the problem of finding positions and values of t errors to solving two systems of linear equations of order t. The Gaussian method can be used for the solution, and then the computational complexity will be of order t^3 [9].

Decoding of the holographic code consists in n^2-fold computation of A_R (i) using formula (4), which is significantly simpler algorithmically and requires less computational resources.

Holographic code is of interest for information transmission over channels with low signal-to-noise ratio (space communication or optical communication systems using free space as a transmission channel, terrestrial, including mobile radio communication, as well as optical communication systems based on RoF technology). In addition, it can improve the reliability of information storage in systems exposed to ionizing radiation (space technology) [12], etc.

2 Increasing the range of atmospheric optical links using holographic coding

Atmospheric optical links utilize the transmission of near-infrared optical radiation through the atmosphere. In English, the corresponding term free-space optics (FSO) also includes transmission in space. FSO systems are used for high-speed communication between two fixed points at distances of up to several kilometers. FSO channels are attractive for a wide range of applications, such as last-mile sections in urban areas, connection of base stations of cellular networks, organization of communication between objects where cable laying is not possible (industrial zones, railroads, etc.), temporary communication channels, communication channels that are not susceptible to external interference and do not create it, reduction of delays compared to cable lines, quantum key distribution, etc.

FSO has a wide range of advantages such as high data rate, information security, low power consumption, and no interference with other wireless communication channels. Atmospheric links with transmission speed of 10 Gbit/s are in commercial use, and experimental links compete with fiber-optic links in terms of speed [13].

However, FSO technology is limited by the effects of atmospheric turbulence and various weather conditions such as rain, haze, smoke, fog, mist, snow, etc.

FSO systems are divided into coherent and incoherent systems based on the method of signal detection. Coherent systems use modulation of the amplitude, frequency or phase of the laser radiation, while incoherent systems use the intensity of the emitted light to transmit information. Coherent systems have much higher sensitivity and resistance to attenuation caused by turbulence, but due to their high complexity and cost, incoherent lines are commonly used in ground-based FSO systems. Various modulation [14] and channel coding [15] techniques are used to improve the robustness of FSO to atmospheric turbulence and fading.

2.1 Modulation and coding in FSO channels

On/off keying (OOK) modulation is the simplest to implement and is often used. When using OOK, the modulated data is represented by the presence ("on") or absence ("off") of a light pulse in each symbol interval. In the receiver, for optimal signal detection, it is necessary to know the instantaneous channel fading factor to set the dynamic threshold [16]. Channel state information can be estimated with reasonably good accuracy using several reference symbols [17].

In addition to the need for dynamic thresholding in the receiver, OOK modulation has relatively low energy and spectral efficiency. Energy efficiency refers to the maximum achievable data rate at the target BER (Bit Error Rate) or minimum BER at the target rate for a given transmission energy. Spectral efficiency refers to the information transmission rate.

To overcome these drawbacks, other modulation schemes have been proposed. A very effective way to solve the problem of energy efficiency is Pulse-Pulse Modulation (PPM) [18]. In the classical method of forming signals with PPM symbol interval [0; T] is divided into L subintervals (slots) with duration t each. The transmission of the l-th symbol corresponds to the transmission of a pulse with time position (l-1)t. It is proved in [19] that for a classical optical channel, PPM almost reaches the channel capacity at limited peak and average power. For signal detection at the receiver, PPM has the advantage that, unlike OOK, it does not require a dynamic threshold for optimal detection [20]. In particular, PPM is proposed for long-range space communications (together with a photon counter in the receiver), where energy efficiency is a critical factor [21].

Multipulse Pulse-Position-Pulse Modulation (MPPM) [22,23], which has advantages over classical PPM in reduced peak-to-average power ratio and improved spectral efficiency [24], has even higher power efficiency. In this approach, not one but K pulses are transmitted during the same L slots. It should be noted that MPPM outperforms PPM also in terms of frequency efficiency [25]. At the same time, MPPM has an increased demodulation complexity [20].

It should be considered that despite the higher bandwidth available in the optical band, spectral efficiency is still an important factor as it is directly related to the required speed of the electronic control, modulation and coding circuitry in the FSO system from a practical point of view. When the peak transmit power is limited, MPPM outperforms PPM. Conversely, when the limitation is imposed on the average transmit power, PPM outperforms MPPM [26].

Another widely used modulation scheme is Pulse-Width Modulation (PWM). Compared to PPM, PWM requires lower peak transmit power, has better spectral efficiency, and is more robust to intersymbol interference, especially for large number of slots per symbol [27]. However, these advantages are counterbalanced by the higher average power requirements of PWM.

A comparative analysis of FSO channel stability to atmospheric turbulence for different modulation methods was carried out in [14]. The study was carried out for OOK modulation, different types of phase modulation (BPSK, 16-PSK, 2-PPM, 16-PPM) and quadrature amplitude modulation (4-QAM, 16-QAM). It is found that 16-PPM position-pulse modulation is the most robust to turbulence and fog in the FSO channel.

The use of orthogonal frequency-domain multiplexing (OFDM) technology in FSO should also be noted. The influence of atmospheric optical channel turbulence on the OFDM signal in different modes has been analyzed in [28-30]. However, when propagating and recovering the transmitted data at the receiving end, due to the inhomogeneity of the medium, the range of the line is significantly limited and the complexity of processing increases. The use of OFDM with discrete Fourier transform and spectrum expansion (DFT-s-OFDM) can be a solution to this problem. DFT precoding moves the input signals into the frequency domain, while DFT-s-OFDM technology can synthesize block single carrier signals with different bandwidths by varying the DFT block size, and also considers the duration of the internal guard interval without affecting the symbol duration, thus realizing lower out-of-band emission (OOVE) compared to OFDM. This technique has flexible implementation mechanisms and lower complexity, which allows controlling the OOVE parameters and the Peak to average power ratios (PAPR) by using, for example, the zero tails or unique words method, which uses a flexible internal guard period instead of a useful data-dependent guard interval. The structure of the DFT-s-OFDM output signal can be interpreted as a precoded OFDM scheme, where the precoding using DFT is aimed at reducing the PAPR. This interpretation has its merits as it can give different precoding strategies to achieve the best parameters. Another feature is that DFT-s-OFDM can be considered as a scheme that increases the sampling of the data symbols by a factor equal to the ratio of the dimensionality of the

IDFT and DFT blocks, and applies cyclic pulse shaping with a sinc function before inserting the guard interval, which is equivalent to signal shaping at a single carrier frequency. In [31], an adaptive modulation method with DFT extension with high spectral efficiency based on frequency multiplexing in a hybrid fiber optic visible optical bandwidth communication channel is presented to reduce PAPR and signal-to-noise ratio.

Despite the use of noise tolerant modulation techniques, fluctuations in received signal intensity caused by atmospheric turbulence can lead to significant performance degradation and system failure. In fact, the atmospheric optical channel has a very long memory and channel fading can cause an abnormally large number of errors that affect thousands of consecutive received channel bits. Reducing fading in FSO channels has been the subject of intensive research. One possible solution is channel coding [32], which is particularly useful for weak turbulence [33]. It is also effective for moderately strong turbulence, provided that the effects of turbulence can first be significantly reduced, e.g., by other fade suppression techniques such as aperture averaging, diversity techniques, or adaptive optics.

A sufficiently large number of works on FSO coded systems have considered the use of convolutional codes, low-density parity check (LDPC) codes [34-36]. The use of LDPC coding together with OFDM modulation is proposed in [37]. Performance bounds on the number of errors for coded FSO communication systems operating in channels with atmospheric turbulence are obtained in [38]. However, these works consider an uncorrelated FSO channel requiring the deployment of large interleavers. An interleaver uses interleaving (mixing) of the transmitted sequence symbols at transmission and restoring its original structure at reception to combat packetizing errors. The coherence time of an atmospheric channel is about 0.1-10 ms, so the fading remains constant over hundreds of thousands to millions of consecutive bits for a typical transmission rate [39]. For atmospheric channels with such large coherence times, this requires high latency and the use of large amounts of memory to store long data frames. In addition, when averaging is used in the receiver aperture, the application of temporal diversity through channel coding becomes more difficult and even impractical [33].

In [40], it is proposed to use feedback channels in FSO to eliminate the need for interleaving and reduce the redundancy introduced with channel coding. The idea is to use the channel state information obtained from the feedback channel to select a suitable encoder-decoder pair from a bank of encoders and decoders [41].

A large number of works on FSO code systems assume the use of binary modulation. But there is also a sufficient number of works in which non-binary modulation is considered. For example, convolutional codes and turbo codes with respect to PPM modulation are considered in [42]. In [17], Reed-Solomon code (RS code) is proposed as a solution for PPM-based modulation. However, RS coding cannot provide satisfactory performance when using hard decoding, which is usually performed at the receiver. Soft RS decoding is computationally too complex and rarely implemented.

2.2 Position holographic coding

The final result of the analysis and selection of modulation and coding methods of radiation in the atmospheric channel is the achievable communication range of the FSO system at a given level of error probability BER. To achieve the optimal result, it is important to ensure the combination of the type of channel coding with the modulation method. An example of such a combination is position-pulse modulation and position immunity codes [43]. One of such codes is a holographic code that uses for transmission not a single pulse in the desired position, as in PPM modulation, but the transmission of a sequence of pulses, which is a one-dimensional linear hologram of a single pulse. The modulation in this case remains positional but becomes multi-pulse modulation, i.e. MPPM. Holographic coding with different characteristics and possible applications is described in [8,12,44].

The peculiarity of the holographic code is that the input coded word must be represented in a single positional code. When using other noise immunity codes, almost all of which are non-position codes, the advantages of PPM (energy efficiency and spectral efficiency) are lost and the potential noise immunity of the selected codes is not achieved.

Let us consider in detail the application of holographic coding in FSO channel with position-pulse modulation.

In the encoding process, the character interval (used to transmit a character, data block) is divided into L slots of duration t. The digital input data block X to be transmitted, which is a k-bit binary code, is converted into a position code A, consisting of $n = 2^k$ points A(i), i = 1,..., n, the value of one of which is 1, the others are zeros: A(i) = 1 at i = X, A(i)=0 at i $\neq$ X. As a result, block A has (n-1) zeros and one unit at the position given by block X (Figure 2.1). Thus, the input data block is used as the address of the unit position in the sequence of zeros of the unit position code. This encoding step corresponds to position-pulse modulation with the number of slots L=n.

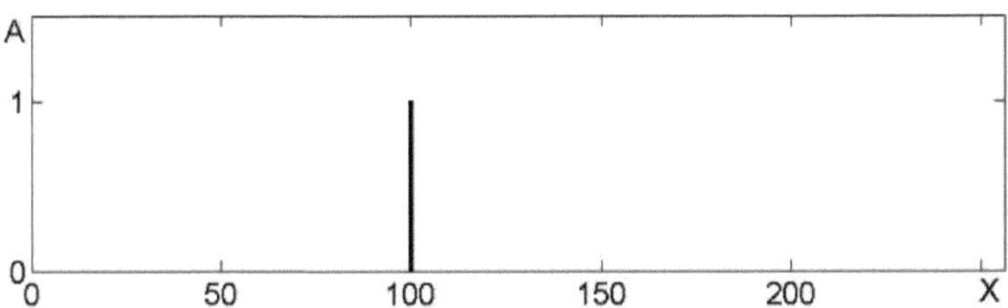

Figure 2.1. Input data block X=100 with PPM modulation at L=256

The holographic coding method is based on mathematical modeling of a one-dimensional hologram created in virtual space by a wave from an object representing an input data block. Formation of the hologram of the position code pulse is performed by constructing a digital sequence corresponding to the Fresnel zone ruler [44]. Thus, a one-dimensional object $A(i)$ is put in correspondence with a one-dimensional hologram $H(j)$. The values of the calculated hologram are rounded to one bit - positive ones are taken as 1, negative ones - as 0. As a result, an n-bit one-dimensional array $H_O(j)$ is formed, which is a code combination with MPPM modulation corresponding to a k-bit input data block X (Fig. 2.2).

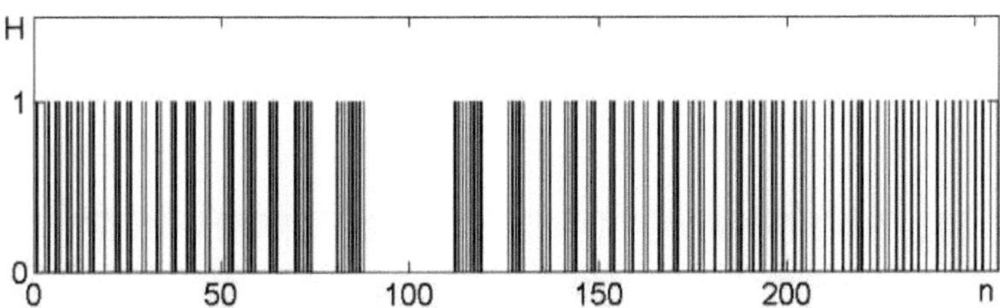

Fig. 2.2. MPPM pulses forming a one-dimensional hologram of the input block X=100 in the codeword with a codeword of diynomial n=256

In the process of transmission during the symbol interval duration T=nt during the slots with numbers corresponding to the single bits of the array H_O (j), the laser is turned on. As a result, a sequence of laser pulses, the number of which is approximately equal to L/2, is emitted into the channel during the symbol interval.

Decoding in the receiver takes place as follows. The signal received during the symbol interval is digitized and subjected to the digital hologram reconstruction procedure. The resulting digital linear array has a maximum in the cell Y with a number corresponding to the value of the input block X. Fig. 2.3 shows the result of modeling in MATLAB environment the process of signal decoding in the presence of white noise in the channel, the power of which is equal to the signal power.

The holographic code utilizes the hologram divisibility property and is able to recover the transmitted information from a hologram fragment as well as the information hidden by noise. Restoration capabilities of the holographic code depend on the hologram length (number of slots L in the symbol interval).

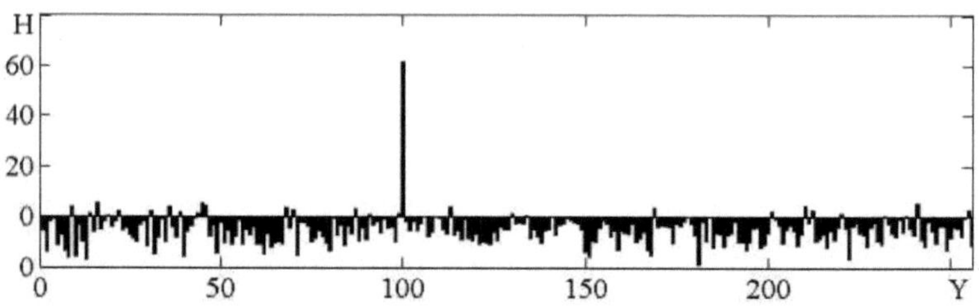

Fig. 2.3. Decoding result at n=256, signal-to-noise ratio S/N=0 dB, decoded value Y=100

In digital communication channels often more informative is an estimation of noise immunity not on a relation signal/noise, and on a limit quantity of random errors in the decoded word. Fig. 2.4 shows the result of decoding of a codeword of length n=256 at presence of 80 random errors (31%). Unambiguously defined maximum in position Y=100 provides correct decoding.

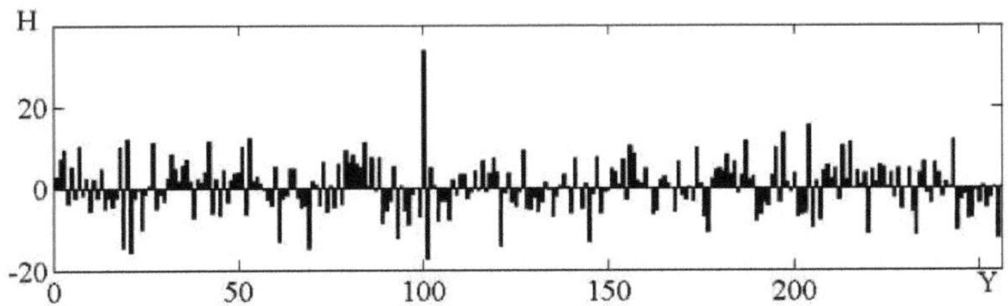

Figure 2.4. Decoding result of the codeword with 31% of random errors, decoded value Y=100

2.3 Atmospheric channel models

There are many FSO channel models depending on the turbulence of the medium and the modulation used. Obviously, the main parameter affecting the signal quality in FSO is attenuation, which has different nature. In [45], devoted to a detailed description of an atmospheric optical communication channel, the receiver input power P_R is related to the transmitted power P_T by the following expression:

$$P_R = P_T \exp(-\tau),$$

where τ is the transmission coefficient. In general case losses in the optical channel are caused by absorption and scattering, which is shown in the formula

$$\gamma(\lambda) = \alpha\,(_m\lambda\,) + \alpha\,(\,) + (\,) + (\,)_{,a}\lambda\,\beta_m\lambda\,\beta_a\lambda$$

where $\alpha\,(_m\lambda\,)$ is the molecular absorption coefficient, $\alpha\,(_a\lambda\,)$ is the aerosol absorption coefficient, $(\beta_m\lambda\,)$ is the molecular scattering coefficient, $(\beta_a\lambda\,)$is the aerosol scattering coefficient.

There are other loss mechanisms related to beam divergence, weather effects: fog, rain, snow, etc. For example, losses caused by fog can be described by the transmission coefficient [14]

$$h_p = \exp(-\sigma z),$$

where the coefficient σ is defined as

$$\sigma = \frac{3.912}{V}\left[\frac{\lambda}{550\text{нм}}\right]^{-q}$$

and the transparency V (in km) and the parameter q are defined in detail in [14]:

$$q = \begin{cases} 1.6, & V > 50 \text{ км} \\ 1.3, & 6 \text{ км} < V < 50 \text{ км} \\ 0.16V + 0.34, & 1 \text{ км} < V < 6 \text{ км} \\ V - 0.5, & 0{,}5 \text{ км} < V < 1 \text{ км} \\ 0, & V < 0.5 \text{ км} \end{cases}$$

Turbulence in an atmospheric channel leads to three main effects [46]:

- if the size of the inhomogeneity is comparable to the wavelength, it leads to a lensing effect;
- if the size of the inhomogeneity is larger than the wavelength, it leads to reflection of radiation;
- if the size of the inhomogeneity is smaller than the wavelength, it leads to scattering of radiation.

The holographic code increases resistance to both signal distortion and loss of signal fragments due to turbulence. The positional representation of coded data in the holographic code facilitates its coupling with positional pulse modulation and makes it possible to transfer the created redundancy necessary to improve noise immunity from the digital channel to the optical channel, replacing PPM with

MPPM. This means that noise immunity is improved without reducing the data rate.

The effectiveness of a particular modulation/coding method is usually evaluated using the bit error rate BER.

Thus, in [14], the following BER estimation technique is proposed for RRM modulation based on Meyer's G-function:

$$P_e(h) = \frac{1}{2\sqrt{\pi}} G_{1,2}^{2,0}\left(\frac{\gamma h^2}{8} \, \bigg|\, {1 \atop 0,\frac{1}{2}} \right),$$

$$BER = QoG_{5,2}^{2,4}\left(\frac{2\gamma h_p^2}{(\alpha\beta)^2} \, \bigg|\, {\frac{1-\alpha}{2},1-\frac{\alpha}{2},\frac{1-\beta}{2},1-\frac{\beta}{2},1 \atop 0,\frac{1}{2}} \right),$$

where P_e (h) is the error probability, h is the transmission coefficient, γ is the signal-to-noise ratio, and the coefficients α and β are determined for the atmospheric channel with gamma-gamma distribution, which is described in detail in [14].

For MRRM, the average BER can be calculated using the formula [46]:

$$\Gamma_e^u \le \frac{M}{8\sqrt{2\pi}} \sum_{j=-Q, j\neq Q}^{Q} H_j \frac{x_j^2 F^2 + x_j w_j F + 2K_n}{(x_j^2 F^2 + x_j w_j F + K_n)^{3/2}} \left(\frac{x_j^2 F^2 + 2x_j w_j F + w_j^2}{w_j} \right) \times$$

$$\times D\sum_{i=1}^{b} d_i \left(\frac{b\mu+\Omega'}{ab} \right)^{\frac{a+i}{2}} G_{1,3}^{2,1}\left[\frac{abK_s}{(b\mu+\Omega')\overline{K}_s} \, \bigg|\, {1 \atop a,i,0} \right]$$

for the M-distribution, also defined through the Meyer G-function, and by the formula

$$P_e^u \le \frac{M}{4\sqrt{2\pi}} \sum_{j=-Q, j\neq Q}^{Q} H_j \frac{x_j^2 F^2 + x_j w_j F + 2K_n}{(x_j^2 F^2 + x_j w_j F + K_n)^{3/2}} \left\{ 1 - \exp\left[-\left(\frac{x_j^2 F + x_j w_j}{\eta \overline{K}_s} \right)^{\beta} \right] \right\}^{\alpha} \left(\frac{x_j^2 F^2 + x_j w_j F + K_n}{w_j} \right)$$

for the exponential Weibull distribution. Note that the BER estimates differ depending on the applied model of turbulence in the atmospheric channel. The symbol error probability (SEP) can be determined for the RRM modulation as follows [47]:

$$p_s(h) = \frac{1}{2}\,erfc(h\gamma_s),$$

where

$$erfc(x) = \frac{1}{\sqrt{\pi}}\,G^{2,0}_{1,2}(x^2\,|^1_{0,1/2}),$$

and in the case of MRRM

$$p_s(h) = 1 - \frac{1}{\sqrt{\pi}}\int\limits_{-\infty}^{\infty}\left[1 - \frac{1}{2}\,erfc(t)\right]^{M-1} e^{-\left(t - \frac{hM\gamma_s}{\sqrt{2}}\right)^2}dt.$$

2.4 Modeling results

To assess the stability of the FSO channel with holographic coding, modeling in MATLAB environment was carried out. Atmospheric turbulence was modeled by superimposing with the help of MATLAB built-in function wgn on the signal H_O (j) of white noise with a given power, as well as by removing a part of the transmitted digital array.

In the process of modeling, the number of digits k (in the range from 4 to 12) of the binary data block transmitted over the FSO channel was set, and thus the number of slots in the symbol interval $L=2^k$. H_O (j) hologram of unit amplitude was synthesized and white noise of adjustable power from -20 dBm to 10 dBm was superimposed on it. Then the reconstruction of the transmitted data block was performed using the noise-distorted hologram and the presence or absence of error in the reconstructed signal was determined. The number of trials was chosen to be sufficient to obtain a stable estimate of the error probability. For comparison, the error probability estimation was performed when transmitting the PPM signal without coding with the same symbol interval length and at the same noise power. Fig. 2.5 shows the dependences of the error probability on the noise power when transmitting information without coding and with coding, obtained for a symbol interval containing L=256 slots.

The data obtained indicate that the use of coding leads to improved noise immunity and reduced error probability in the receiver and has an effect equivalent to that of increasing transmitter power.

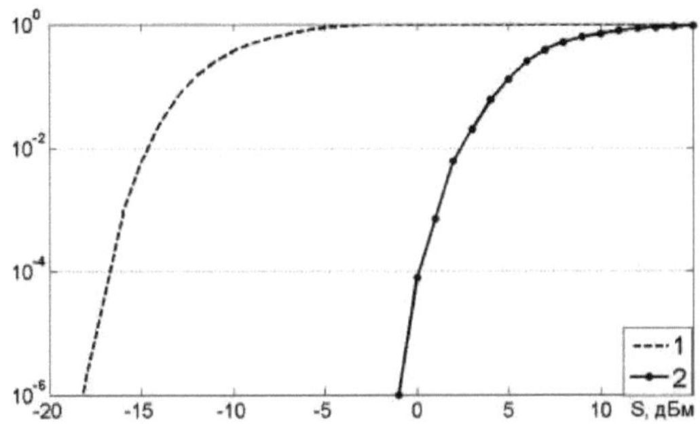

Fig. 2.5. Dependence of the error probability P on the noise power S at L=256. 1 - without coding, 2 - with coding

Fig. 2.6 shows the dependence of the generated equivalent power reserve of the transmitter on the length of the character interval, determined by the number of digits of the input data block, with a constant error probability equal to 10^{-3} . From the graph shows that at k = 9 (character interval is divided into 512 slots) is formed 100-fold power reserve. This reserve can be used both to increase the communication range, and to improve the reliability of the communication channel. For example, at L=256 slots in the symbol interval in the absence of coding to ensure an error probability of no more than 10^{-6} noise power in the channel should be no more than -18 dBm. When coding is used, the error probability of 10^{-6} is provided at a noise power of -1 dBm, i.e. 50 times higher.

The great advantage of positional coding is the fact that the information redundancy required for noise-resistant coding is created inside the symbol interval and does not change the symbol sequence frequency, i.e. does not reduce the speed of information transmission. In all non-position codes introduction of redundancy leads to reduction of code speed.

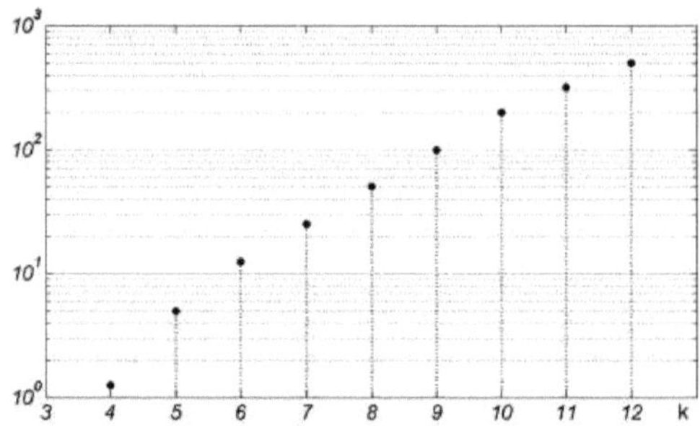

Fig. 2.6. Dependence of the equivalent power margin on the
number of digits k of the transmitted data block

When evaluating the possibility of using holographic coding in existing FSO systems it is necessary to take into account that for hologram transmission multi-pulse position-pulse modulation is used and, accordingly, in the symbol interval is transmitted not one pulse (as in the widely used PPM), but L/2 pulses. This increases the average transmitter power by L/2 times.

Thus, the considered method of digital information transmission via atmospheric channel is based on the joint use of position-pulse modulation and position holographic coding. In this case, the coding takes place inside the symbol interval and can be considered as a component of multi-pulse position-pulse modulation MPPM. Accordingly, we can speak of this scheme as holographic MPPM - HMPPM.

If in the used FSO hardware there is no possibility of transition to MPPM, holographic coding can be applied to the information arriving at the FSO input. In this case, information redundancy is introduced before FSO, as a result, more data is transmitted over the atmospheric channel and the rate of transmission of useful information is reduced. But all the advantages in noise immunity and power boosting equivalent remain.

The main problem preventing the range increase of open laser communication lines is signal attenuation in atmospheric turbulence. Repeated attempts to change the situation with the help of noise-resistant coding methods have not given a significant effect. One of the reasons is the lack of consistency between the chosen methods of modulation of optical radiation and channel coding methods. Combining two homogeneous operations - position modulation and position holographic coding into one method of holographic multi-pulse position-pulse modulation gives a significant gain in noise immunity and increased communication range of atmospheric optical communication links.

3 Spectral holographic coding

In modern optical research, problems related to registration, processing and analysis of amplitude and phase characteristics of time-varying optical signals of arbitrary structure are set and solved on the basis of new approaches. One of the methods of solving these problems is the introduction of additional spectral modulation [48]. Since we are talking about amplitude and phase information, the solution of such problems for time-varying signals is usually carried out using holographic methods, including spectral holography [49].

Spectral modulation is applicable, for example, in wireless information transmission systems using ultra-wideband noise-like signals. It gives high noise immunity, electromagnetic compatibility and the ability to transmit information messages "deep below the noise" in communication channels with strong interference [50].

3.1 Spectral coding

Combining two methods - holographic coding and direct processing of the signal spectrum in the method of spectral holographic coding gives even higher result in increasing noise immunity of information transmission systems.

In [51] the possibility of coded spectral modulation of a stream of binary information symbols and coherent compression of broadband noise-like signals in the receiver as a result of double spectral processing with subsequent restoration of the transmitted information was shown.

GMSK (Gaussian Minimum Shift Keying) modulation is widely used in digital radio communication systems, such as GSM. For GMSK, one of two modulation schemes is usually used - frequency-controlled oscillator or quadrature modulator. Modulators/demodulators can be either analog or digital, but in either case there are transients associated with frequency establishment as well as filter group delays that limit the minimum duration of the pulse train.

At spectral holographic coding with discrete frequency manipulation the coding process consists in forming N radio pulses at orthogonal frequencies (the set of frequencies is defined by the holographic coder for each parcel) by switching (manipulation) of the outputs of N harmonic signal generators working in constant mode. Therefore, there are no transients associated with frequency changes, as there are no filters of any kind. Formation of the group signal occurs by summing the manipulated harmonics at the input of the transmitter amplifier or directly in the antenna.

At decoding it is necessary to carry out detection in the received signal each of N used frequencies. In analog technology, this requires N selective filters with a very narrow bandwidth. In digital processing it is necessary to digitize the received signal (without filtering), perform a discrete Fourier transform and determine in the received parcel the presence or absence of each of the N frequencies. The resulting binary sequence (1 - presence of harmonic, 0 - absence), representing the spectrum of the signal, is subjected to holographic decoding and the transmitted binary word is extracted from it.

Spectral holographic coding with discrete frequency manipulation provides gains in noise immunity due to the combined action of three mechanisms: frequency manipulation, spectral coding, and holographic coding.

In [44], holographic noise-tolerant coding applied to a signal to be transmitted over a digital communication channel is described. A binary k-bit block of source data is replaced by an N-bit ($N=2^k$) codeword representing a linear one-dimensional hologram of a virtual point source whose position in space is determined by the value of the coded block.

However, due to the duality of signal representation, similar transformations are also possible in the frequency domain, applicable to the signal spectrum. Depending on the transmission conditions and characteristics of the communication channel, this can give a gain in noise immunity. One of such methods is spectral holographic coding.

In contrast to the method described in [44] for coding a time-varying signal by replacing its shape by the shape of its virtual hologram, spectral holographic coding creates a signal with a given spectrum. The shape of the spectrum is the same one-dimensional hologram whose values are rounded to one bit and represents a binary sequence (Fig. 3.1).

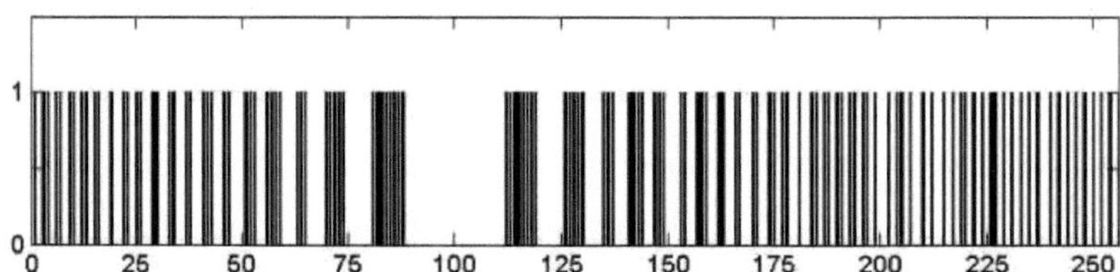

Fig. 3.1. Spectrum shape - linear hologram (sequence of zeros and ones) of length N=256

The coded block value is the position number of the hologram center (center of Fresnel zones) in the codeword, in this case X=100.

To create a signal with a spectrum of this form it is enough to add a set of harmonics of equal amplitude with numbers corresponding to the numbers of positions of units in the hologram. Harmonics with numbers corresponding to the positions of zeros do not participate in the formation of the signal. This operation is a discrete frequency manipulation of N harmonics.

The issues of minimizing the error of information transmission using multi-pulse position-pulse modulation over laser communication channels are considered in [52]. This modulation method has common features with holographic coding, as in both cases position coding is used.

If all harmonics have the same initial phase, the total signal contains a large emission at the beginning and end (Fig. 3.2), which is undesirable for energy reasons. Therefore, the phases of the harmonics should be distributed, for example, according to a random law. The signal in this case has a noise-like form (Fig. 3.3).

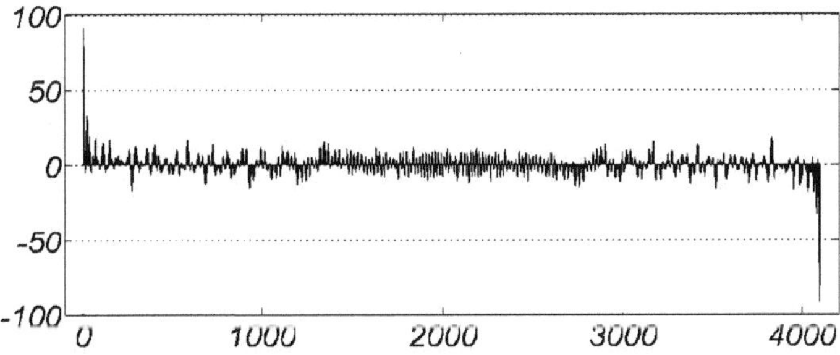

Fig. 3.2. Signal formed by the sum of in-phase harmonics

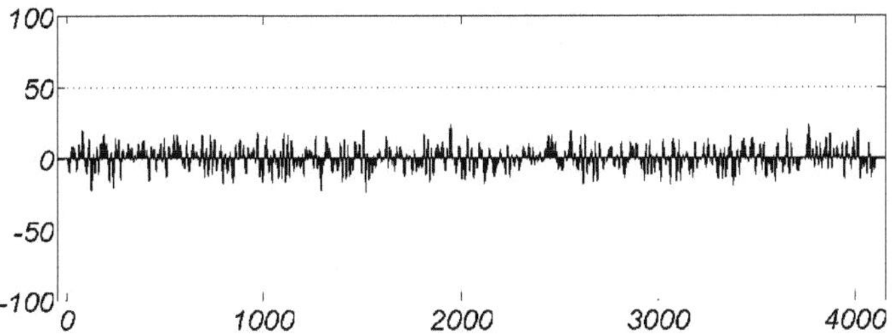

Figure 3.3. Signal formed by the sum of harmonics with random phase

An important requirement is that the duration of the signal (codeword) must be equal to the whole number of periods of all the harmonics used (in this example - one period of the lowest harmonic and 256 periods of the highest). In this case, the spectrum of the signal is discrete (linear) and strictly corresponds to Fig. 3.1.

When additive white noise is superimposed on the signal, the signal spectrum is also distorted (Fig. 3.4).

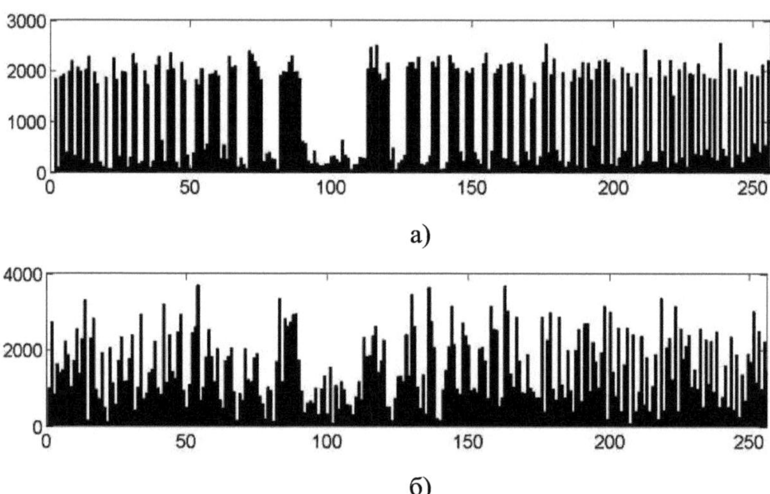

a)

б)

Fig. 3.4. Signal spectrum at the receiver input at signal-to-noise ratio +5 dB (a) and -5 dB (b)

3.2 Modeling results

The spectral holographic code has been studied by modeling in MATLAB environment the process of coding and transmission of information in the presence of white Gaussian noise.

Signal decoding in the receiver was performed by calculating the spectrum of the received signal, and the inverse holographic transformation was applied to it, which restores the position number of the hologram center - the value of the original data block (Fig. 3.5).

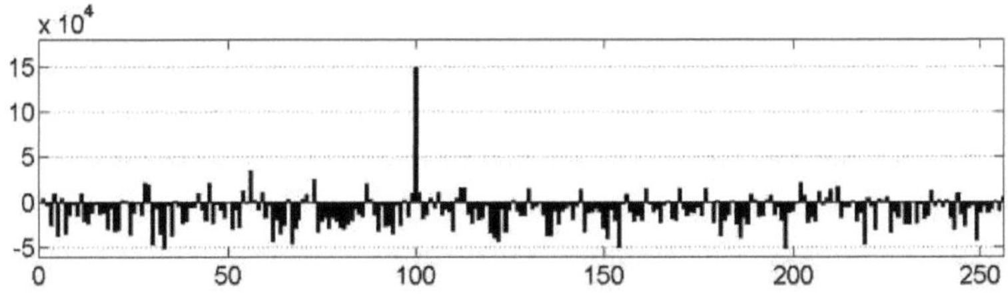

Fig. 3.5. Result of hologram restoration at the ratio
signal-to-noise -10 dB

The key factor determining the noise immunity of the method under consideration is the relative duration of the total harmonic signal. In the example under consideration the frequency of the highest harmonic is 16 times less than the sampling frequency of the signal in the receiver. Thus probability of error of decoding makes 0,01 at a ratio of signal/noise -18 dB. In Fig. 3.6 shows the graphs of signal and noise in one scale for this case.

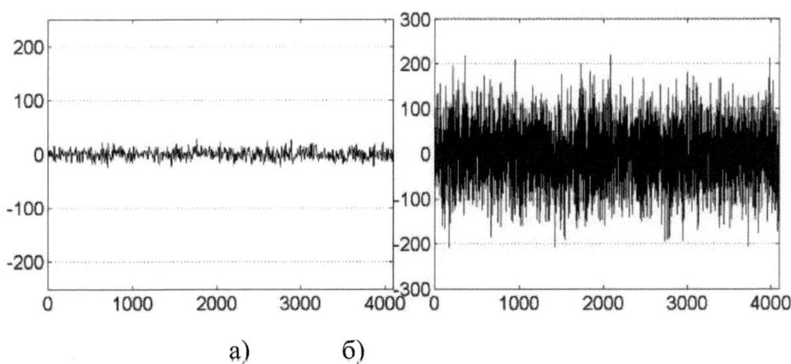

a)   б)

Fig. 3.6. Signal (a) and noise (b) at signal-to-noise ratio -18 dB

Increasing the number of samples in the period of the higher harmonic (increasing the sampling frequency) increases the gain in noise immunity.

For estimation of noise immunity of spectral holographic coding the comparative modeling of process of information transmission on a channel with additive white Gaussian noise at use of widely applied codes is carried out. Dependence of decoding error probability on signal/noise ratio in a channel for Reed-Solomon code (RS-code), Reed-Maller code (RM-code), majoritarian code, holographic and spectral codes is considered. For this purpose, the results obtained in [44] are added to the results obtained in [44] to simulate the performance of the spectral code. In all cases the number of digits of the source word is 8, the length of the codeword is 256 bits (code rate R=1/32). The results are shown in Fig. 3.7.

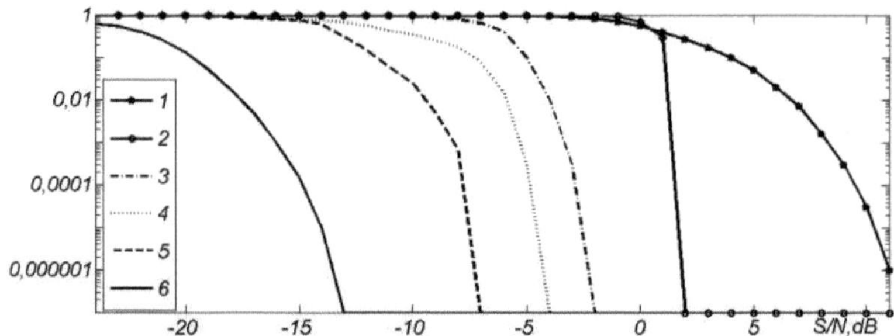

Fig. 3.7. Dependence of decoding error probability P_o on signal-to-noise ratio at code rate R=1/32: 1 - without coding, 2 - RS code, 3 - RM code, 4 - majority code, 5 - holographic code, 6 - spectral code

The graphs show that spectral holographic coding provides a gain in noise immunity of 7-8 dB compared to coding the signal directly.

Another advantage of the spectral code is less complexity of encoding and decoding when the redundancy is varied over a wide range, and high secrecy due to the use of noise-like signal.

The spectral holographic code is of interest for information transmission over low signal level channels, such as in image transmission over long-distance space communication channels, where the task of receiving a weak signal at the receiver's thermal noise level has a higher priority than the transmission rate.

3.3 Use of spectral approach in image and arbitrary data processing

Holographic methods are increasingly used in information processing, including both the use of holograms themselves and methods of processing digital arrays that are not images. At the same time all kinds of digital holograms form large volumes of data, especially if they capture large viewing angles and objects with significant depth. And if holographic data are to be transmitted or stored, means of information encoding and compression become of high importance.

There are a large number of holographic image compression standards, including Jpeg, Jpeg2000, Vp9, and HEVC/H.265 [53]. In many cases, image compression methods exploit the similarity between different parts of the image in the time and frequency domains and preserve fragments of similar data to reduce the size of the final file with minimal damage to quality. In addition, state-of-the-art compression methods use various mathematical operations for error correction. In [54], a hybrid HEVC-Wavelet compression algorithm is presented, which utilizes the HEVC central kernel and a 2D-Wavelet channel to predict the compression error.

In [55], the state-of-the-art of holographic data coding is reviewed and a variant of the HEVC method based on directional transform is proposed to improve compression and coding efficiency. In [56] combinations of methods consisting of frequency filtering of the hologram, splitting the Fourier spectrum of the filtered hologram into real/magnitude and amplitude-phase parts, obtaining its wavelet decomposition by different transformations and additional processing of wavelet coefficients are investigated.

For effective information processing, besides the task of hologram compression, it is necessary to ensure their reliable transmission over communication channels and/or storage. The fact that the internal information redundancy of holograms can be used both for compression and for increasing noise immunity in adjustable ratios provides interesting possibilities in this respect.

In [44] the use of holographic coding of arbitrary digital data to improve noise immunity and reliability of information transmission is considered. A further extension of the field of application of holographic methods of information processing is the transition to the spectral domain, spectral holography. Direct processing of the spatial spectrum of an image is of interest for filtering, modulation and coding tasks. One of the methods for solving these problems is the introduction of additional spectral modulation [48]. Since amplitude and phase information are involved, the solution of such problems for time-varying signals is usually performed using holographic methods, including spectral holography [49]. Spectral modulation is applicable, for example, in wireless information transmission systems using ultra-wideband (UWB) noise-like signals [57].

In [50], it is proposed to modulate the spectrum of a digital information sequence representing a UWB noise-like signal and recover the information in the receiver using spectral processing.

In [44], holographic noise-immune coding in the space of a virtual object, the information carrier, is described. However, due to the duality of information representation, similar transformations are also possible in the frequency domain, applicable to the spatial spectrum of the object. Combining two methods - holographic coding and direct spectrum processing in the method of spectral holographic coding gives even higher result in increasing noise immunity of transmission and storage systems, both images and arbitrary information.

The factor expanding the scope of possible applications of holographic coding is the emergence of tensor holography [58]. A convolutional neural network capable of synthesizing a hologram in a memory of less than 1 MB will be able to provide real-time holographic encoding/decoding.

When digital images are transmitted over communication channels, the image is converted into a linear array by line-by-line scanning. During transmission, this array is affected by both broadband noise and narrowband interference. To protect against noise, noise-resistant coding that requires redundancy (e.g., holographic coding) is used. To eliminate narrowband noise, digital filtering is usually used, but in some cases it is more appropriate to use not the filtering of the distorted image, and processing directly its spectrum - spectral filtering.

In [51], image filtering operations in the frequency domain using filters of different profiles are considered. The filter performs multiplication of the spatial spectrum of the image by a spectral window expressed by the function $W(\omega)$. The subsequent inverse Fourier transform generates the processed image. Adaptation algorithms can be used to construct the spectral window function [59]. Highly efficient filters for noise filtering, frequency correction and impulse noise filtering can be realized on neural networks [60]. However, adaptation and neural network algorithms, while having high efficiency, require large computational resources. At the same time, direct processing of the spatial spectrum of an image is realized with much lower costs.

The spectral approach allows increasing the efficiency of not only filtering, but also compression of both images and holograms. Data compression can be performed in two ways - with or without loss of information. A distinctive feature of images, and moreover holograms, as a form of information representation, is the presence of large internal redundancy. It allows to use compression methods with information loss, if losses do not exceed permissible limits.

Despite the availability of a large number of image compression methods [61-65], the task of finding more efficient compression methods with lower computational complexity remains relevant. The transition from the image space to the spatial frequency domain may provide such a possibility.

When considering signals used as image carriers in lossy compression algorithms to select an optimal set of algorithm parameters for loss minimization, it is necessary to take into account the characteristics of images both in the spatial and frequency domains. For this purpose, frequency decomposition (Fourier, wavelet, etc.) or geometric interpretation can be used [62]. One of such approaches, compression of phase holograms using deep neural network training, is discussed in [66].

High efficiency with adjustable level of insertion distortion is provided by the widely used JPEG compression method, which uses discrete cosine transform (DCT) realized by matrix

$$DCT\text{-}2_n = [\cos(k(l+1/2)\pi/n)]_{0\leq k,l<n}\,.$$

For this variant of fixed dimension vector DDC, there are algorithms to minimize the number of multiplication operations.

The use of wavelet transforms (WT) allows to obtain a higher compression ratio by removing small details of the image. Of great importance is the choice of a method for spatial-frequency partitioning of the EP spectrum, for example, by further decomposition of high-frequency sub-bands to obtain an optimal basis [63]. To improve the decomposition efficiency, the adaptation of the basis to the image content with quantitative estimation of the signal entropy by the basis of wavelet packets y_k [64] can be used:

$$H = -\sum_k y_k^2 \ln(y_k^2).$$

Image compression using DDC and EP requires a rather large amount of additional mathematical operations. At the same time, there is a possibility to reduce computational costs by processing the spectrum of spatial frequencies of images. Besides, it is possible to additionally increase the efficiency of systems of formation and transmission of digital images through communication channels due to the complex use of compression and noise-resistant coding algorithms [67-69].

Another method that can be used to improve the efficiency of noise-tolerant image coding and compression is compressed sensing (CS). It is a method for efficient information gathering and recovery by finding solutions for an underdetermined system, based on the principle that sparse function recovery requires fewer samples than required by the sampling theorem [70,71]. Strictly speaking, the CS method provides lossy recovery, so the widest area of its applications is related to image processing. At the same time, holographic coding of arbitrary digital information is accompanied by the introduction of redundancy, which makes it possible to apply CS also to the processing of digital signals that do not allow lossy information. Thus, the methods of holographic coding and spectral compression of images can be considered as belonging to the class of detection methods with compression.

3.4 Spectral coding of arbitrary digital information

In the method of holographic noise-immune coding described in [44], a binary k-bit block of source data is replaced by an N-bit ($N=2^k$) codeword representing a linear one-dimensional hologram of a virtual point source whose position in the virtual space is determined by the value of the coded block. Unlike this method, which uses instead of the data block (source object) its virtual hologram, in spectral holographic coding a function with a given spectrum is created. The shape of the spectrum in this case is the same one-dimensional hologram, the values of which are rounded to one bit and represents a sequence of zeros and ones - a one in the i-th position means the presence of the i-th harmonic in the spectrum, zero - the absence.

To create a function y(mT) with such a linear spectrum it is enough to add a set of harmonics of equal amplitude with numbers corresponding to the numbers of unit positions in the hologram:

$$y(mT) = \sum_{i=1}^{N}(G(i)\cdot\sin(2\pi(m/M)\cdot i + r(i)\cdot 2\pi)),$$

where G(i) is the linear array of the hologram, N is the number of harmonics, M is the number of samples in the signal, r(i) is a random number in the range 0...1.

Harmonics with numbers corresponding to the positions of zeros do not participate in the formation of the function y(mT). This operation is a type of orthogonal frequency-division multiplexing (OFDM), which is characterized by the fact that the frequencies of N orthogonal subcarriers are in multiples, and amplitude manipulation is used as digital modulation.

As a result, a function with a linear spectrum representing a hologram of the initial object is synthesized (Fig. 3.8).

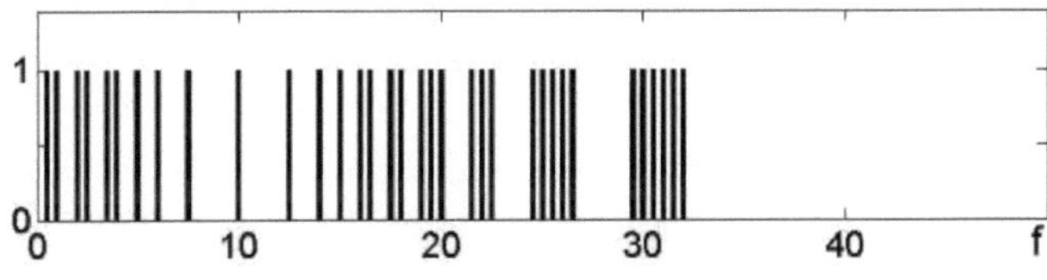

Figure 3.8. Spectrum - hologram

The procedure of synthesis of the function y(mT) can be hardware implemented in analog form, if the integer ratio of all frequencies to the first harmonic is rigidly maintained. However, in many cases it is more accurate and simpler to perform digital synthesis, which can be implemented in two ways. The first method is the algebraic addition of N harmonics forming the spectrum-hologram, and the second method is the inverse fast Fourier transform of the synthesized spectrum.

There is one requirement for the duration of the function y(mT) - it must be at least one period of the first harmonic. The duration of a larger value does not affect the noise immunity of coding, but proportionally increases the amount of processed information.

47

An important characteristic is the peak factor of the function y(mT), which reaches a maximum at in-phase harmonics. The higher value of the peak factor imposes higher requirements on the linearity of the amplifier to avoid increasing out-of-band emissions and reducing the noise immunity of the communication channel [72-74]. The best result is achieved when the harmonic phases are distributed according to a random law. The function y(mT) in this case has a noise-like form.

To perform the inverse transformation and restore the object by the function y(mT), its spectrum is calculated. In this case, to construct the spectrum, a duration normalized fragment is used, containing an integer number of periods of each harmonic - one period of the first harmonic, two periods of the second harmonic, and so on up to N periods of the harmonic with the number N. Fulfillment of this requirement allows to obtain a linear spectrum containing no side harmonics. This requirement limits the duration of the function y(mT) from below, but does not impose restrictions from above.

The digital array representing the spectrum is treated as a one-dimensional hologram of the original digital object, and is decoded using the holographic method described in [44].

Transfer of holographic coding from the object space to the spatial frequency domain gives an additional gain in noise immunity. To estimate the noise immunity of spectral holographic coding, a comparative modeling of the process of superposition of additive white Gaussian noise using widely used codes has been carried out. The choice of the image sampling period in the presence of noise should be carried out taking into account not only the characteristics of the image to be recorded, but also the noise level [75].

In Fig. 3.9 shows dependences of decoding error probability Po on signal-to-noise ratio for Reed-Solomon code (RS-code), Reed-Maller code (RM-code), majoritarian code and holographic code obtained in [44], supplemented by modeling results for spectral code. The modeling was carried out for an 8-bit word of the initial data at a codeword length of 256 bits (code rate R=1/32).

The graphs show that holographic coding in the spatial spectrum region provides a gain in noise immunity of 7-8 dB compared to coding in the object space.

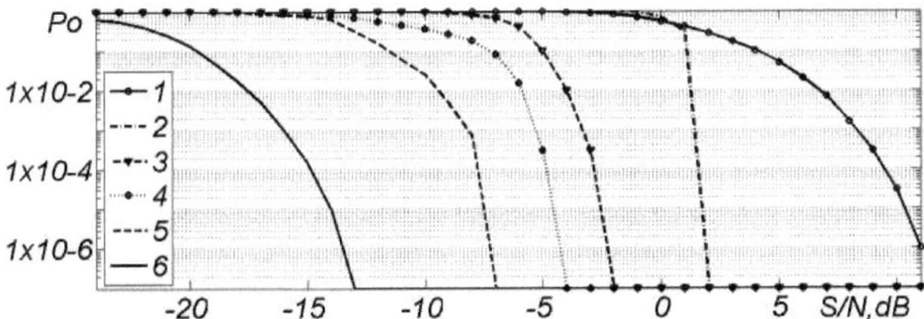

Fig. 3.9. Dependence of Po decoding error probability on signal-to-noise ratio: 1 - without coding, 2 - PC-code, 3 - RM-code, 4 - majoritarian code, 5 - holographic code, 6 - spectral code

Another advantage of the spectral code is less computational complexity for a wide range of code rates.

Spectral holographic code can be used, for example, in image transmission over long-distance space communication channels, when the task of error-free reception of a weak signal at the receiver's thermal noise level has a higher priority than the transmission rate.

3.5 Spectral image processing

Operations with the spatial spectrum of images allows you to perform filtering and image correction (blurring, sharpening, brightness and contrast adjustment, etc.), as well as algorithmically more complex operations, such as image compression.

The simplest way to realize the filtering of narrowband interference is to perform a discrete Fourier transform, in the resulting digital spectrum zero out the frequencies where the interference is present and perform the inverse Fourier transform. If the problem is solved by digital filters, if it is necessary to remove several harmonics in different parts of the spectrum, the total order of filters increases proportionally, but the computational cost to realize spectral filtering remains the same - direct and inverse Fourier transform.

Spectral compression of both holograms and images is more convenient to evaluate on the example of images. Consider a test image containing rather large areas with smooth brightness change and a large number of small details (Fig. 3.10).

The study of the image and its spectrum, modeling of spectral compression was carried out in MATLAB environment.

Fig. 3.10. Original image

The spatial spectrum of the test image is shown in Fig. 3.11. For further processing we need a digital spectrum in the frequency band from 0 to the sampling frequency f_d , so the spectrum in Fig. 3.11 contains the main and inverse copies.

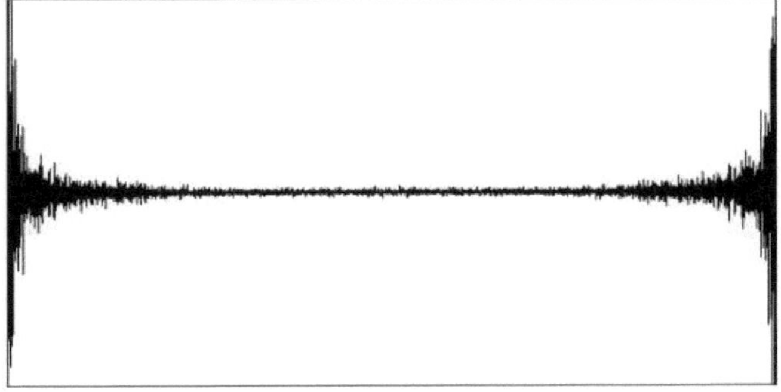

Fig. 3.11. Spatial spectrum

An obvious way to reduce the amount of recorded information is to erase the least significant part of the spectrum (in this case, the high-frequency part), e.g., as shown in Figure 3.12, erasing the high-frequency half of the spectrum in the frequency range from f /4 to f /2. 3.12, deleting the high-frequency half of the spectrum in the frequency range from f_d /4 to f_d /2. All operations performed in the main part of the spectrum should be mirrored in the inverse copy, i.e. in this part the frequencies from f_d /2 to 3f /4 are deleted.$_d$

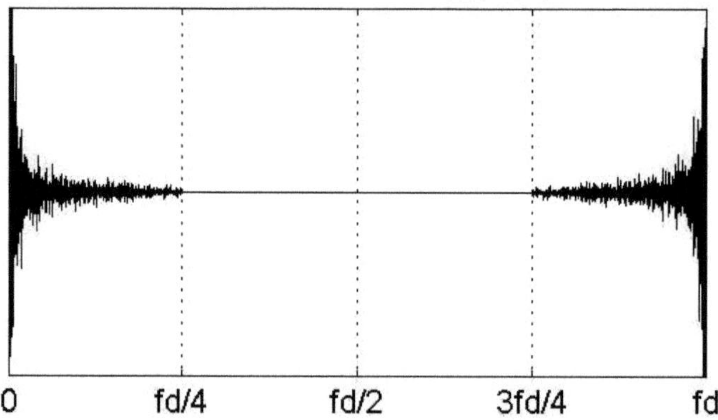

Fig. 3.12. Removal of the high-frequency part of the spectrum

A more detailed consideration of the spectrum structure can provide additional opportunities to reduce the amount of information. At sufficiently large magnification it is noticeable that the spectrum has a periodic structure with the number of elements equal to the number of lines in the image (Fig. 3.13).

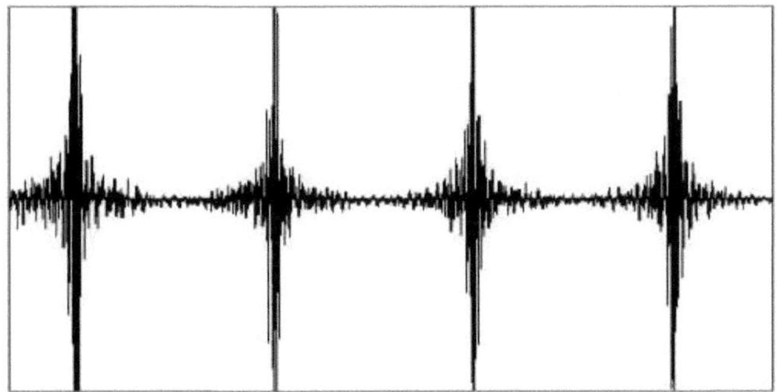

Fig. 3.13. Periodic structure of the spectrum

In each element of this structure, the center part is low level, has little effect on the quality of the full image and can therefore be reduced - in Fig. 3.14 half of each period is removed.

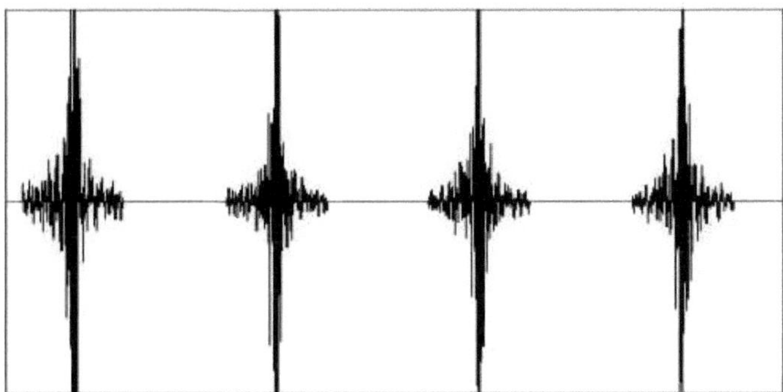

Fig. 3.14. Deletions within periodic spectrum fragments

Thus, the removal of the high-frequency half of the full spectrum and the central half of each periodic element gives a total reduction in the amount of information in the image by a factor of 4. At the same time, the image quality decreases insignificantly (Fig. 3.15).

To form a compressed image file it is necessary after removing spectrum fragments to compact the remaining fragments and record them as a single array. Thus, the compressed image format is a thinned and compacted spectrum of the original image. To restore the image it is necessary to move the spectrum fragments to their original places and perform the inverse Fourier transform.

Fig. 3.15. 4 times compressed image

When transmitting images and arbitrary digital data over communication channels, noise-tolerant coding for protection against broadband noise and narrowband interference is of great importance. Spectral approach to coding and compression of information allows to reduce the requirements to computational resources and use the gain from reduction of the volume of transmitted information to introduce redundancy in noise-resistant coding, for example, by holographic code, and significantly increase the reliability of information transmission.

4 Transmission of digital information in the form of images over multimode fiber optics

The capacity of fiber optic communication lines is continuously growing. In 2020, a speed of 178 Tbit/s will be achieved for single-mode fiber for digital information transmission [76]. At the same time, in optics, image transmission occurs at an incomparably higher speed - a complete image is carried in space by a wavefront for one period of a light wave $(2 \times 10^{-15}$ s).

The idea of transmitting an image over a multimode fiber emerged shortly after the advent of fiber optics [77]. This use of a multimode waveguide requires giving it the ability to distribute image information between individual modes at the input of the waveguide in a controlled manner, and then extract it in an orderly manner at the output. Multimode optical fibers offer great promise for increasing data transmission capacity, especially for applications such as optical communication systems [78], fiber lasers [79], and endoscopic imaging [80-84]. However, phase distortions (delays) in light propagation prevent the preservation of image details [85]. The biggest problem for image transmission through a multimode fiber is its mode dispersion, since phase (time) delays lead to significant disruption of the transmitted image, which starts already at short distances even in ideal straight fibers [86].

The purpose of this work is to increase the transmission rate of digital information in the form of images over multimode fiber by providing higher resistance to mode dispersion.

4.1 Methods of image transmission over multimode fiber

For multimode stepped fibers, there is a formula for approximate calculation of the number of modes [87]

$$M \approx V^2/2, \quad (1)$$

where V is the normalized frequency, defined as

$$V = 2\pi r_1 \, NA/\lambda , \quad (2)$$

where NA is the numerical aperture of the fiber, r_1 is the radius of the fiber core, λ is the wavelength of radiation. In [87], a model of image propagation in a multimode fiber is described for the case when the distance between the imaged object and the fiber aperture is several orders of magnitude greater than λ, and an assumption is made about the weakly guiding fiber approximation when the difference between the refractive index of the core n_1 and the cladding n_2 is less than 1% - such an approximation allows us to pass to the formalism of linearly polarized (LP) modes. In this case, we can assume that only paraxial rays propagate through the fiber, and the field propagating along the fiber can be described as

$$E(r,\phi,z) = \begin{cases} \sum_{n,m} a_{nm} \psi_{nm}(r,\phi) \exp(-j\beta_{nm}z), & r \leq r_1 \\ 0, & r > r_1 \end{cases} \quad (3)$$

where

$$\psi_{nm}(r,\phi) = \frac{J_n(k_{nm}r)\exp(jn\phi)}{N_{nm}^{1/2}} \quad (4)$$

The modes of the fiber are described by Bessel functions, n and m represent azimuthal and radial order, respectively; N_{nm} is the normalizing integral; a_{nm} is the mode weighting factor depending on the spatial distribution of the incident radiation, β_{nm} is the propagation constant, different for each mode, causing the phase difference between the modes, which is the cause of mode dispersion. As mentioned above, in the weakly guided fiber approximation, we can consider LP modes obtained by a combination of transverse electric (TE), transverse magnetic (TM) and hybrid electric (HE) modes having the same propagation constant β - the modes are said to be degenerate. Such modes satisfy the general dispersion relation obtained by simplifying the characteristic equation at $n_1 \approx n_2$:

$$\frac{uJ_m(u)}{J_{m+1}(u)} = -\frac{\varpi K_m(\varpi)}{K_{m+1}(\varpi)},\tag{5}$$

where $J_m(u)$ and $K_m(\varpi)$ are Bessel functions of the first and second kind of order m, respectively, describing the mode field in the core and shell, and the parameters u and ϖ are defined as

$$u = r_1\sqrt{(k_0 n_1)^2 - \beta^2},$$
$$\varpi = r_1\sqrt{\beta^2 - (k_0 n_2)^2},\tag{6}$$

where $k_0 = 2\pi/\lambda$ is the wave number. In this case, the degeneracy of modes implies a strong coupling of modes within the same mode group (e.g., the LP_{01} mode is formed by the GE_{11} and PE_{01}), but the mixing of modes between different groups (e.g., between LP_{01} and LP_{11}) will be much smaller. Nevertheless, in quartz multimode fibers, strong mode coupling will be observed already at distances of the order of 300 m [88]. Strong mode coupling means reduction of mode dispersion, which is very relevant in the framework of the considered problem.

In [84], an algorithm for calculating and correcting the mode dispersion based on the separation and equalization of the contribution of each mode was proposed. Based on simulations, it is concluded that the algorithm works with both stepped and smooth refractive index fibers, as well as with fibers of any length. In [89], the problem of improving the efficiency of input of spectral power of radiation into the fiber is considered and it is proposed to encode the spectrum to form a broadband signal, which is equivalent to replacing wavelength multiplexing with code-division multiplexing. In [90], attempts made to circumvent the limitations imposed by mode dispersion are analyzed and three approaches are highlighted. The first one is compensation of phase difference of different modes by changing the waveguide structure, the second one is equalization of phase velocity by creating a special structure of electromagnetic field transmitted through the waveguide and the third one is using special image coding at the waveguide input and reverse decoding at the output. It is noted that these methods did not give significant progress in solving the problem. At the same time it is shown that modes of the waveguide are eigenfunctions of the Fourier transform and it is proposed to transmit through the waveguide not the image itself, but its Fourier spectrum formed by the lens. The fact that the Fourier components of the electromagnetic field when propagating along the waveguide under conditions of forced paraxiality cannot diverge in space leads to the independence of the resolving power of the system from the length of the waveguide. As a result, [90] concluded that the transition to the spatial spectrum can allow transmitting images to an unlimited distance, and special diffractive optical elements can be added to equalize phase velocities and eliminate intermode interactions (in the case of fiber deformations) [91].

The main disadvantage of using an imaging optical element, such as a lens, is that pre-calibration of the optical system is required. A database of calibration input/output pairs, which is stored digitally, is used to generate the fiber output image. If any changes occur in the optical system after the calibration data has been collected, the system parameters deteriorate rapidly. In particular, fiber bending is a major limiting factor for such systems [92]. Several works have proposed another approach to approximate the unknown transmission function of a multimode fiber, which consists of sending a large amount of input data (not necessarily orthogonal) over the degrees of freedom of the system and collecting the output data in order to learn the transmission properties using a deep learning neural network [93-97]. Perhaps most importantly, drift of system characteristics (such as bending) can be incorporated into the training set and lead to improved performance reliability. In [94], image visualization of temperature and mechanical fluctuations of a 1 km long fiber using a neural network was demonstrated.

Nevertheless, despite numerous attempts made in this area, the problem of qualitative transmission of sufficiently complex images over fiber has not been solved to date. At the same time, another possibility of using the limited potential of multimode fiber as a transmission channel (relatively small number of paraxial modes, high level of mode dispersion and other distortions) remains insufficiently investigated - the possibility of increasing the speed of transmission of arbitrary digital information via fiber.

4.2   Transmission of digital information

One way to increase the transmission rate is to increase the number of parallel channels using Spatial Division Multiplexing (SDM) [98]. This method uses the transverse spatial extent of a multimode fiber to create parallel data channels. However, acceptable link quality with an acceptable level of errors due to mode dispersion and fiber losses is obtained only at short distances. In [99] the transmission of signals with spatial multiplexing in the optical fiber up to 1 km long was considered and confirmed experimentally, but for this purpose it was necessary to use a small-mode fiber with three spatial streams. This solution may well be used to organize an optical distribution network in buildings. In [100, 101] the methods of intermode interference reduction are considered, which allow increasing the number of modes in small-mode fiber up to 10 and more.

Despite the successes in this direction, the task of developing more effective methods of using multimode fiber for high-speed transmission of information remains relevant. Let us consider the technology of image transmission over multimode fiber for transmission of digital information of arbitrary type. The obvious solution is to convert the digital array to be transmitted into an image. However, on this way there are difficulties due to the difference of criteria used to assess the quality of image transmission intended for visual perception and transmission of arbitrary digital array for subsequent machine processing. In the first case, sufficiently large distortions are allowed and quality assessment is subjective, while in the second case, bit-by-bit coincidence of the transmitted and received array is necessary. In the conditions of multimode transmission in the transmitted signal distortions are introduced by mode dispersion, intermode interference (arising, for example, at fiber deformations), waveguide, which plays the role of spatial frequency filter, image rastering due to discreteness of the Fourier spectrum and other sources. Therefore, a special form of representation of the digital array as an image is required.

To increase the stability of the transmitted information to errors and distortions occurring during transmission, it is of interest to transmit along the fiber not the image corresponding to the input digital array, but its hologram, and to use the specific property of holography - hologram divisibility, which provides the restoration of the original image by a highly distorted hologram. The use of hologram to improve the noise immunity of atmospheric optical communication lines is described in

[102]. However, in these cases the information is transmitted over a serial communication channel and therefore a one-dimensional array of source data and a one-dimensional hologram are used. To use a hologram as a form of image representation, the hologram must be two-dimensional and serve as an encoded image of the digital array to be transmitted.

The possibilities of using holograms in the process of image transmission over multimode fiber have been investigated in [103-108]. In [103] it is shown that a set of individual holograms, each of which produces a single target spot, can be combined into a complex superposition for fast generation of multiple spots. In [104], a system is described that dynamically compensates with a digital hologram for beam pointing instability, external perturbations introducing low-order aberrations, and fluctuations in the relative phase of the modes supported by the fiber. In [105-107], the problem arising in holographic optical tweezers, where phase holograms for a single trap are easily computed, is considered and it is shown that for multiple traps, an intensity distribution close to the superposition of the intensities of the individual holograms can be obtained [108].

In all these cases, the hologram carries information about the original image embedded in the comparative brightness of the image points. Therefore, the transmitted hologram is affected by mode dispersion, intermode interference, temperature and mechanical fiber fluctuations to the same extent as the original image. To increase the stability of the transmitted information to errors caused by all these factors, it was proposed in [44] to use a single position code rather than a binary one to represent the initial block of digital information. In this case, the optical object for which the hologram is constructed is a point source on a black background, and the information is embedded in the coordinates of a point on the object field. The brightness of the object does not matter, so one bit (0 or 1) is enough to specify it. The result of encoding is the simplest hologram - a Fresnel zone plate, the coordinates of the center of which carry the encoded information, and the image of which is transmitted over the fiber. In the receiver the purpose of decoding is not to restore the brightness of each point of the transmitted hologram, but to calculate the coordinates of the center of Fresnel zones. Exactly it causes high stability of this method to all kinds of distortions of the transmitted image.

Let us consider the possibility of transmission over a multimode fiber of block-by-block digital information, where each block corresponds to the image of a zone plate, the coordinates of the center of which are set by the transmitted block. For example, a 10-bit data block is divided into two 5-bit coordinates of a bright point on a black source image of size 32x32, which has 1024 pixels. The center of the Fresnel zones (of the transmitted hologram) has the same coordinates. Obviously, the number of hologram pixels should not exceed the number of modes in the fiber.

The hologram image is transmitted along the fiber by one of the considered methods of multimode transmission and is subjected to all types of distortions inherent in the chosen method.

To restore the value of the original array on the receiving side, the hologram obtained at the fiber output must be subjected to inverse transformation. This can be done optically by creating an interference pattern in the plane of the photodetector array and determining the coordinates of the brightest point. The operation of restoration of the initial array can also be done digitally, processing the image of the received hologram fixed by the photodetector matrix. Small size of the transmitted hologram (not more than 1 kbit - an array of 32x32 single-bit points) allows to do it with simple hardware resources.

4.3 Modeling results

In the process of hologram transmission over multimode fiber it is subjected to phase distortions due to the presence of mode dispersion, in addition, there are distortions due to the inherent sensitivity of the fiber to thermal, acoustic, mechanical perturbations. All these reasons eventually lead to errors in the transmitted digital array - inversion of single-bit pixels of the received digital hologram. Accordingly, the most complete criterion of the quality of digital information transmission is the bitwise comparison of the transmitted and received digital array, evaluated by the dependence of the probability of error recovery of the received array on the intensity of bit errors in the transmission channel.

Modeling of digital information transmission over multimode fiber is carried out in MATLAB environment. The realized algorithm contains the following operations:

- Converting the input data block into an image (virtual object)
- Building a hologram of the object
- Introduction of random errors due to mode dispersion and other causes into the digital hologram
- Digital reconstruction of an object by hologram
- Converts an object image into an output digital data block.

The input data block is a 10-bit binary word, which is used to set the X,Y coordinates of the single point with level "1" on the image of the object of size 32x32, all other points of which have zero value (Fig. 4.1a).

Object hologram - image of a zone plate, on which the center of Fresnel zones has X,Y coordinates of a single point of the object (Fig. 4.1b).

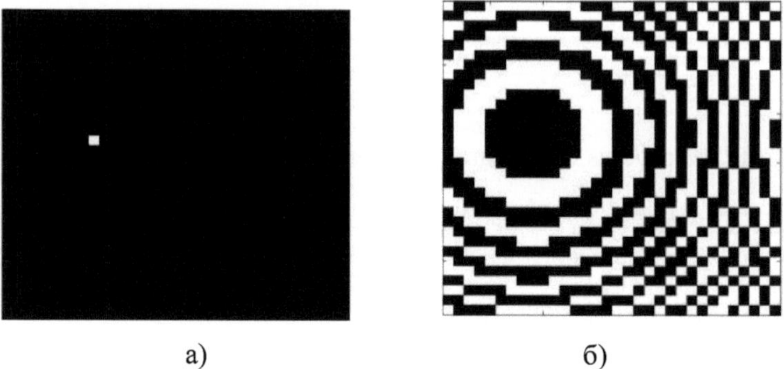

a)   б)

Fig. 4.1. Image and hologram of the object.

Image of the object (a), hologram of the object (b)

The result of restoring the image of the object by hologram (brightness function in the plane of formation of the restored image) in the absence of distortions is shown in Fig. 4.2.

The reconstructed image of the object, as can be seen in Fig. 4.2, contains background noise due to the finiteness of the number of elements of the digital hologram. This effect is also manifested on the hologram itself (Fig. 4.1b) - an analog hologram would consist of concentric rings. But the presence of noise does not prevent the accurate determination of the coordinates of the brightest point in the reconstructed image of the object and allows to do it with a large margin of noise immunity.

When simulating the process of image transmission along the fiber, the transmitted information was distorted by random errors. The hologram of the object containing 30% of errors is shown in Fig. 4.3.

The result of restoration of the object image from the distorted hologram is shown in Fig. 4.4.

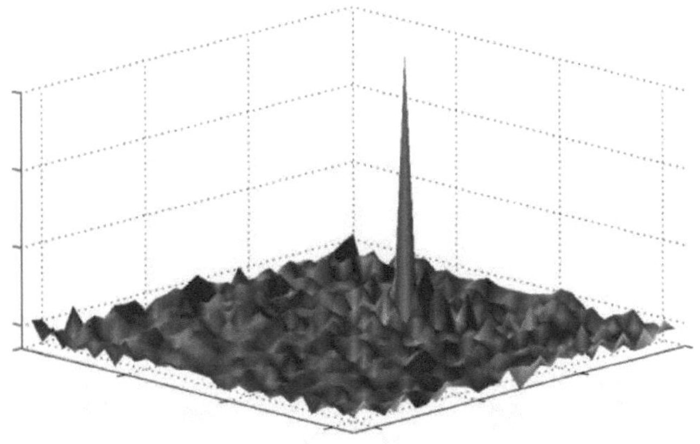

Figure 4.2. Brightness function R(X, Y) of the reconstructed object

Figure 4.3. Distorted hologram

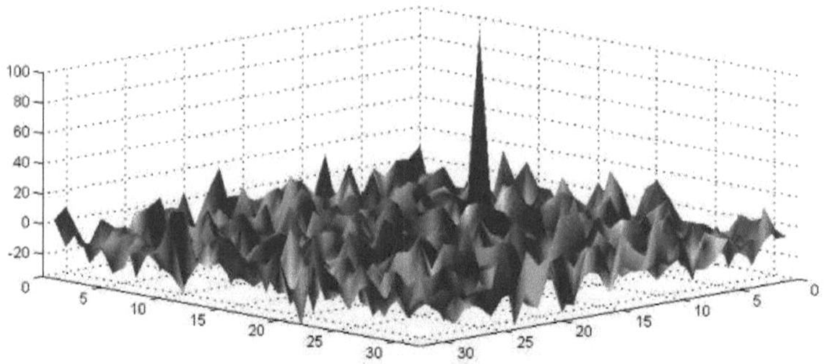

Fig. 4.4. Brightness function R(X, Y) of the object reconstructed from a
distorted hologram

As Fig. 4.4, the distortion of 30% of the hologram area, despite the growth of the noise level, does not prevent the unambiguous determination of the coordinates of the brightest point, and, therefore, does not lead to distortion of the transmitted information.

The dependence of the probability of error-free recovery of the input data block on the degree of hologram distortion was investigated. The number of trials was chosen sufficient to obtain a stable estimate of the error probability. It was found that at the hologram size of 32x32 the probability of correct recovery is 0.9997 in the presence of 40% of errors in the hologram.

The volume of transmitted information is determined by the number of bits of the input data block and in the considered variant is 10 bits. The amount of information transmitted by one image can be increased by introducing several data blocks when constructing the image of the object. Fig. 4.5a shows the image of four data blocks, in Fig. 4.5b is their hologram, and Fig. 4.6 is the result of the reconstruction.

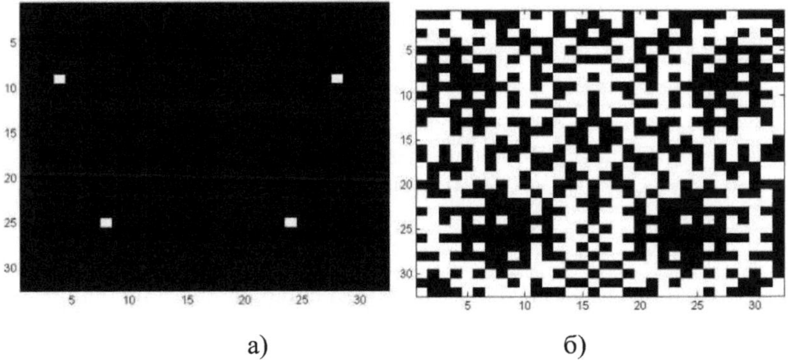

a)   б)

Fig. 4.5. Image of an object and its hologram. An object consisting of four data blocks (a), a hologram of an object consisting of four data blocks (b)

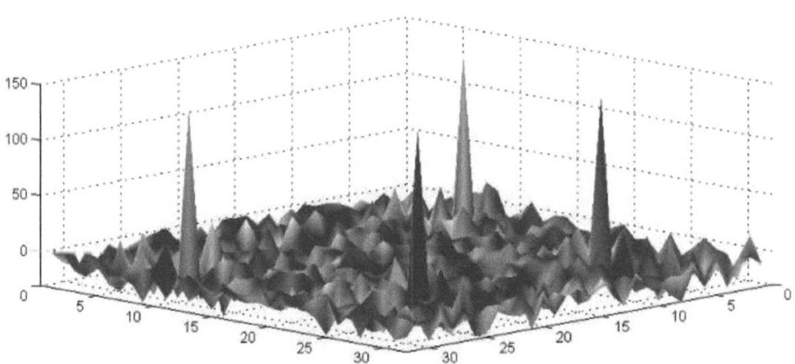

Figure 4.6. Brightness function R(X, Y) of four object image data blocks

As can be seen in Fig. 5.6, the transmission of four data blocks is accompanied by an increase in the noise level, but the noise immunity margin still leaves the possibility of suppressing mode dispersion and other distortions.

Thus, increasing the amount of information contained in one image leads to an increase in the speed of information transmission. When transmitting information in the form of a 32x32 hologram using one block of data for one clock, 10 bits are transmitted, and using four blocks - 40 bits. It means increase of speed of information transmission over multimode fiber in 10 and 40 times in comparison with usual transmission over digital channel. The choice of optimal combination of hologram size (number of used modes), number of digital blocks transmitted in one image and noise immunity margin, determining the maximum transmission range, depends on the application area and the task at hand.

Currently, many works are published in the field of image transmission over multimode fiber. However, the developed methods do not allow to realize the transmission of complex images with acceptable quality. At the same time, the limited capabilities of these methods can be successfully used to transmit arbitrary digital information. Unlike real images, which require high spatial resolution, digital information can be transmitted as small images ranging in size from 8x8 to 32x32. The problem of mode dispersion and other types of distortions arising in fiber transmission is solved by using instead of the object image its hologram, which due to the inherent property of holography divisibility is highly resistant to distortions. As a result, the transmission range increases significantly at a controlled level of probability of error of the transmitted message and the speed of information transmission increases by 10-40 times.

5 Increasing the active utilization period of on-board electronic equipment of spacecrafts

For electronic equipment of space systems, and first of all memory devices, the problem of protection from the effects of ionizing cosmic radiation and other external factors that distort the stored and processed information is relevant [109]. Radiation effects and space particles create a large number of errors accumulating in memory devices. The use of known methods of noise-tolerant information coding has an effect for a limited time until the number of errors becomes too large. Responsible systems use ECC memory (error-correcting code memory) - a type of computer memory that automatically recognizes and corrects spontaneous changes (errors) of memory bits - one error in one machine word. When the length of a machine word is 64 bits, the number of corrected errors is < 1.5%.

To improve the reliability of information storage, a form of data recording that provides restoration of a block of information by its fragment - the holographic method of recording using the property of hologram divisibility (the possibility of restoring the full image of an object by a hologram fragment) is of interest [110].

5.1 Holographic method of information recovery

The idea of using holographic coding principles was formulated in [111,112], but full digital modeling of a hologram required large computational resources, so they considered pseudoholographic coding. According to the proposed method, the elements of a digital two-dimensional array are uniformly mixed in a certain way, as a result of which a reduced copy of the original array can be reconstructed from any part of the reordered array. The study of pseudoholographic methods is continued in [113-116]. The described methods have an area of application limited to the problems of coding arrays of information with large internal redundancy, and are analogous to the iterleaving method [117] used in communication systems to combat packet errors.

The use of full holographic coding for error correction was proposed in [44]. The considered method is based on modeling the hologram as an interference pattern of a flat image formed by a matrix representation of the initial digital data block. Encoding and decoding operations in this case require rather large computational resources. However the complexity of calculations can be considerably reduced if to take into account that for a digital hologram the number of points rather than their mutual arrangement has a determining value. In [44] it is shown that the coding efficiency is preserved at transition from a matrix hologram to a linear hologram with the same number of points. Therefore, it is rational to use one-dimensional data arrays and one-dimensional holograms.

The use of the holographic method of information conversion to increase resistance to ionizing radiation of information processing and storage systems was proposed in [12,118]. Let us consider the possibility and efficiency of using the holographic method of noise-resistant coding in memory devices exposed to external factors leading to random and deterministic (packet) errors.

5.2 Modeling results

The study of the corrective ability of the holographic code is carried out by modeling in MATLAB environment the process of distortion of the hologram H_O by random and packet errors.

Figure 7.1 shows a view of a linear hologram of an 8-bit input data block with the value X=99. In this case the size of the hologram written into memory is 256 bits, redundancy factor is 32.

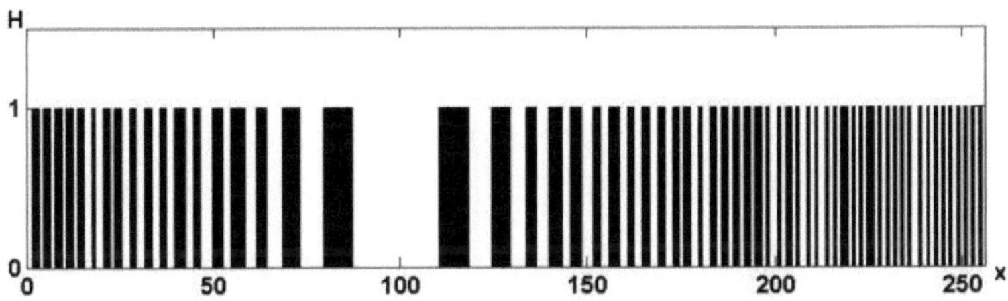

Figure 7.1. Hologram H_O for X=99

Figure 7.2 shows the result of decoding A $_{,R}$ in which the position of maximum Y=99 carries information about the encoded value. In the received array there is a small noise of decoding, caused by a finite number of discrete values of the hologram, and does not prevent to allocate the information value.

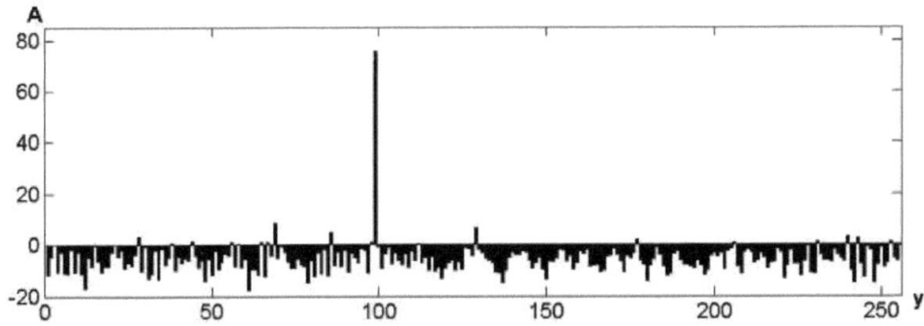

Figure 7.2. Reconstructed array A_R at n=256, Y=99

Let's consider the robustness of the code to erasures, random and packet errors.

View of a hologram at the decoder input at erasure (loss) of 75% of a hologram of size n=256 is shown in figure 7.3.

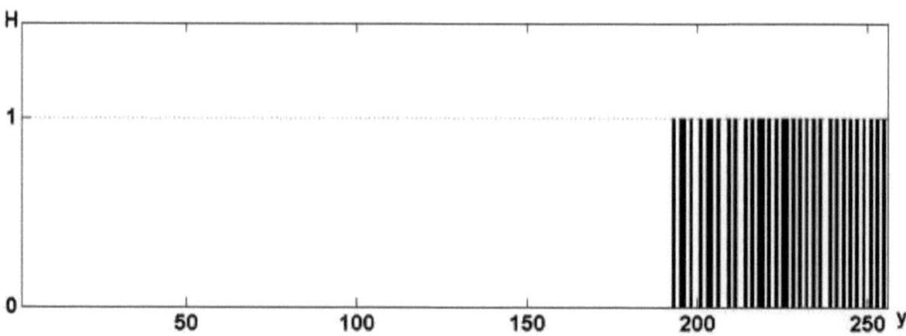

Figure 3. Hologram of H_R for X=99. Losses of 75%

The result of data block recovery by the remaining 25% is shown in Figure 7.4. The maximum point in position Y=99 corresponds to the transmitted value X=99 and unambiguously determines the value of the encoded block.

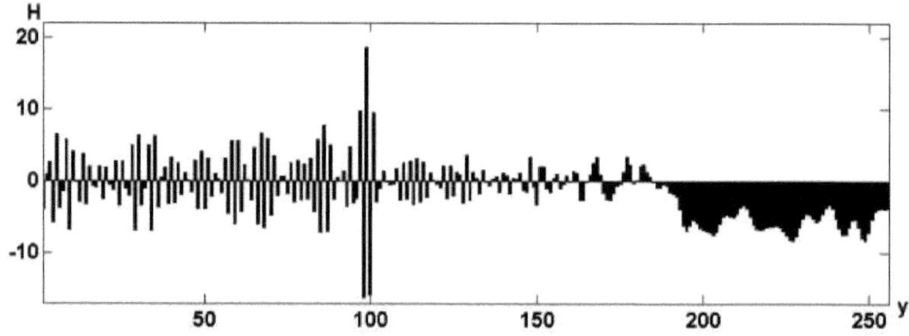

Figure 7.4. Reconstructed array A_R at 75% loss, n=256, Y=99=X

Holographic coding provides resistance not only to information loss but also to random errors. The occurrence of errors is modeled by replacing a part of the hologram with a binary random sequence (noise). The array reconstructed from a hologram of size n=256 containing 75% of noise is shown in Figure 7.5.

Increasing the hologram size leads to increasing noise immunity. At $n=2^{14}=16394$ successful information recovery occurs at the length of the noise sequence up to 95% of the hologram size.

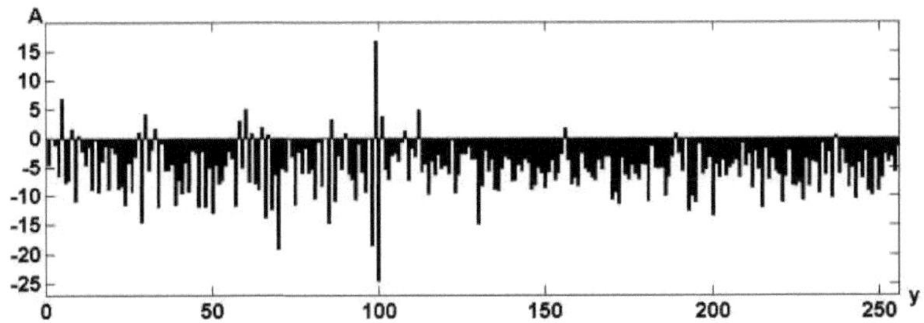

Figure 7.5. Reconstructed array A_R at noise sequence length 75%, n=256, Y=99=X

When replacing a part of the hologram with binary noise, the number of errors occurring is less than the number of noise positions, since about half of the noise positions will coincide with the information bits and will not create errors. Therefore, the maximum possible number of independent random errors in a sufficiently large hologram is 50% of the number of bits in the hologram. If the number of errors is more than 50%, the errors are dependent, and 100% of errors corresponds to a completely deterministic case - bitwise inversion of the hologram.

The most difficult situation for decoding is the number of random errors approaching to 50%. Corrective ability of the holographic code depends on the hologram size n. Statistics of modeling results shows that at n=256 probability of decoding error makes 10^{-3} at number of errors at the decoder input 30%. At number of errors 25% and number of tests 10 000 decoding errors are not fixed. At n=1024 probability of decoding error 10^{-3} is reached at 41% of errors in the hologram.

73

The considered examples of information recovery are characteristic for cases with independent random errors. At the same time, most binary information systems are characterized by correlation between errors and their combination into packets [119]. Packet errors occur under intensive exposure to ionizing radiation, when recording information on data carriers, as well as in communication channels [120].

To correct errors occurring during information storage, noise-resistant codes are widely used. One of the most effective is Reed-Solomon code (RS-code), which is widely used in systems of data recovery from CDs, in creation of archives with information for recovery in case of damage, in noise-resistant coding [121]. The limit of corrective ability of PC-code is defined by Singleton's bound [122], according to which for error correction the code should have at least two check symbols per error. At a large degree of redundancy the number of correctable errors approaches 50% of the codeword length. The peculiarity of the PC-code is that it demonstrates such a high correcting ability only for packet errors [119], yielding, for example, to the Reed-Muller code (RM-code) in correcting independent random errors. RM-code with codeword length $n=2^m$ corrects $2^{m-2}-1$ errors of any kind [122], occupying almost 25% of the code combination.

The basic decoder of a holographic code, as well as RS-code, eliminates errors occupying not more than 50% of a codeword. However by virtue of specificity of holographic method of information representation it is possible to construct the universal decoder correcting any quantity of grouping packet errors up to 100% of hologram size, i.e. when all symbols are distorted.

In itself the task of correction of 100% errors is trivial - for this purpose it is enough to invert each bit of the codeword. However here for all codes, except holographic, there is a problem of a choice of one of two equally probable results of decoding - direct or inverted. The holographic code gives the identical result of decoding, both for the undistorted codeword, and for the codeword containing 100 % of errors. Figure 7.6 shows the result of decoding of the inverted block containing 100% of errors. From here we can see that packet errors lead to inversion of the maximum in the recovered array, but its position is preserved, so the information recovery is correct.

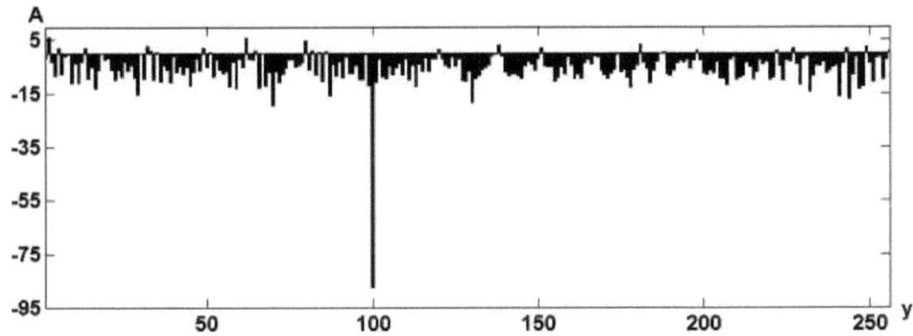

Figure 7.6. Reconstructed array A_R (Y=100), number of errors - 256 (100%)

Recovery is more difficult at packet error rate of about 50%. To solve this problem the decoded data block is divided into two equal parts and each part is decoded in direct and inverted form. Each of the four decoding variants forms a complete output array, with all implementations having different noise levels (Figure 7.7). Joint analysis of these arrays allows to determine the value of the decoded data block.

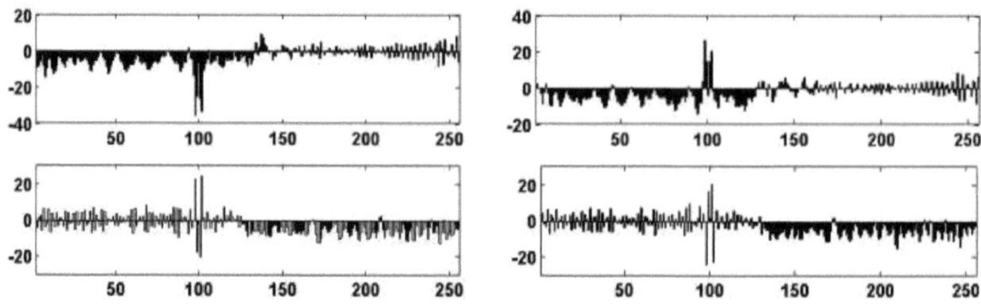

Figure 7.7. Results of work of four decoders

Figure 7.8 shows a fragment of one of four arrays at decoding of the hologram containing 128 errors at n=256 (50% of errors), encoded value Y=100. It shows that, despite the absence of an extremum at the point Y=100, the value of the input block can be recovered by a characteristic combination of symmetrically arranged four lateral maxima.

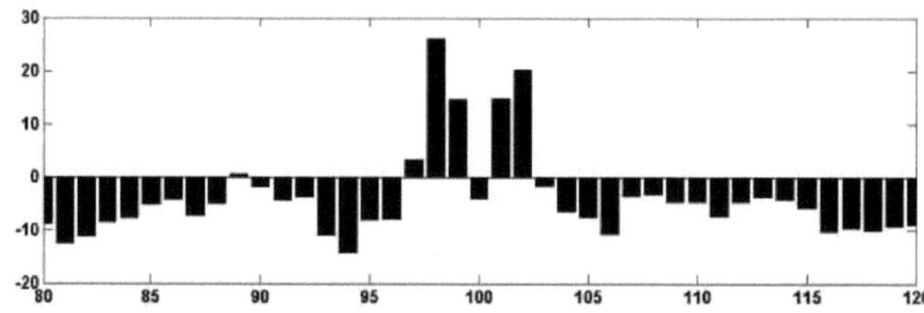

Figure 7.8. Histogram fragment of the reconstructed array A_R (Y=100)

Simulations have shown that a universal decoder containing 4 decoders and a maximum selection block corrects any number of packet errors - from 0 to 100% of the recorded data block. This decoder is effective at elimination of random and packet errors. At number of errors less than 50% they are random and the decoder provides information recovery at 41% of errors in the codeword. At number of errors more than 50% they are dependent and form the right part of the graph (Figure 7.9).

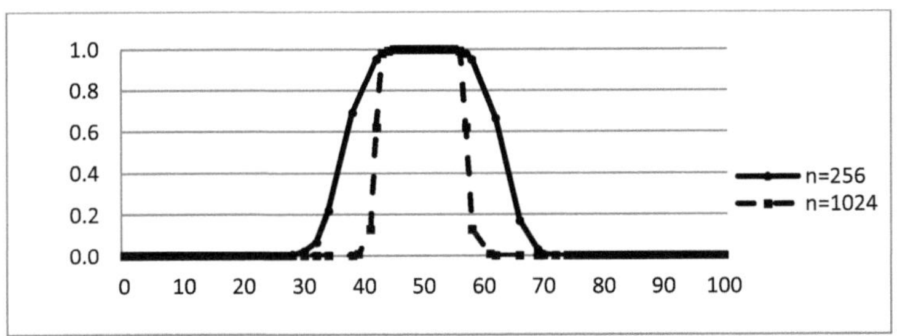

Fig. 7.9. Probability of an error at the decoder output depending on number of errors at the decoder input for holograms of sizes n=256 and n=1024

Thus, holographic coding corrects errors if their number is less than 40 or more than 60 percent of the codeword length at n=1024 (Figure 7.9). This allows to increase reliability of data recovery in information storage systems exposed to ionizing radiation, temperature and other factors causing degradation of element base parameters.

The information storage system resistant to ionizing radiation is a memory array of increased capacity taking into account the selected redundancy factor, and a memory controller that performs holographic coding when writing information and decoding with automatic error correction when reading information. The algorithm of the controller itself can be implemented in the form of a programmable logic integrated circuit or stored in a permanent memory device that is not affected by ionizing radiation.

It can be used in any computer architecture. For this purpose, it is necessary to modify the memory controller by installing an encoding/decoding module in it.

6 Improving the positioning accuracy of the GLONASS system

One of the main characteristics of the global navigation satellite system GLONASS is the accuracy of coordinates and altitude estimation obtained in the navigation equipment of consumers (NAP) only by satellite signals without additional information [123]. The positioning accuracy is of particular importance for navigation systems of aircraft, including unmanned ones, in conditions of unintentional and intentional interference [124,125]. However, the insufficient positioning accuracy in many cases requires the development of other ways to solve this problem. In [126] it is proposed to use for solving fundamental geodynamic and geodetic problems a system of high-precision determination of ephemeris and time corrections, which collects, stores and processes measurement and navigation information on GLONASS satellite navigation systems. In [127] computational methods of reducing errors caused by signal passage in the ionosphere and troposphere are proposed, in [128] it is shown that the use of spatial selection methods with the help of antenna array allows to significantly improve the accuracy of navigation determinations by reducing the influence of multipath reception. In [129], the idea of digital registration of navigation signals is described and a method of high-speed post-processing is proposed to improve the accuracy of estimates. An approach to improving the noise immunity of navigation equipment by using a scheme of deep complementation of navigation equipment is proposed in [130,131]. A method to improve the noise immunity of positioning systems by transmitting an additional timestamp is considered in [132]. The most complicated and costly is the creation of a new navigation-information satellite system proposed in [133], which eliminates the problems in the development of the GLONASS system. The characteristics of the NAP have a great influence on the positioning accuracy. Standard samples of NAPs do not provide the necessary level of noise immunity at the existing power level of received GLONASS signals of the order of minus 166...156 dBW [134, 135].

The accuracy of coordinate measurement is determined by the number of satellites simultaneously visible to the navigation equipment. GLONASS errors are 3-6 m when 7-8 satellites are used. Up to 11 GLONASS satellites are simultaneously above the horizon on most of the earth's surface, but the signal-to-noise ratio in the communication channel, necessary for error-free reception of information, is often provided only for 2-4 satellites. Figure 8.1 shows an example of visibility of satellites of different navigation systems in urban areas.

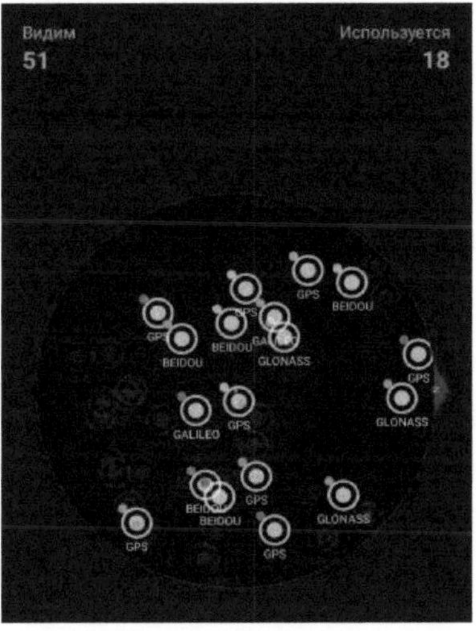

Fig. 8.1. Visibility of GPS, GLONASS, BEIDOU, GALILEO satellites

Figure 8.2 shows the signal levels for visible satellites.

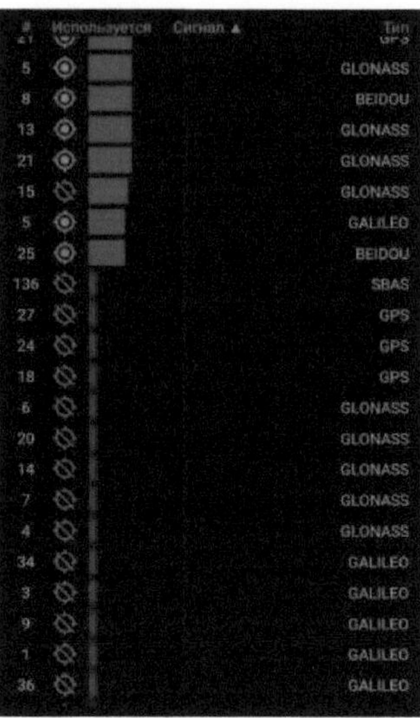

Figure 8.2. Visible and used satellites

The required signal-to-noise ratio is maintained for 4 GLONASS satellites, 5 GLONASS satellites are in the line of sight but are not used due to low signal-to-noise ratio. Under these conditions, the positioning error exceeds 10 m.

In the GLONASS system to improve noise immunity is used Hamming code that corrects single errors. Transmitted blocks of digital information is an 85-bit code, in which the upper 77 digits contain information symbols, and the lower 8 digits - verification [134]. The positioning error can be reduced not only by increasing the number of satellites in the satellite constellation, but also by using coding that provides correction of a larger number of errors.

To date, a large number of correction codes of different efficiency have been developed. However, their efficiency is insufficient to restore the original message when transmitting information via satellite communication channel in conditions of low signal-to-noise ratio, when large fragments of information may be lost. One of the ways to improve the stability of the communication channel is to use the form of signal representation, which provides the restoration of the message on its fragment. Such a feature possesses the holographic method of corrective coding [44]. The process of coding the transmitted information is a mathematical modeling of a digital hologram created in the virtual space by a wave from the encoded object. Holographic coding can be used to improve noise immunity of communication channels, reliability of information storage, as well as to improve the accuracy of the GLONASS system.

6.1 Holographic encoding/decoding algorithm

At the first stage, the transmitted digital message, which is an n-bit binary code, is converted into a unit position code with the number of positions $N=2^n$. By this, redundancy with the coefficient $q=2^n/n$ is put into the message. The block in the unit code has (N-1) zeros and one unit in the position given by the original message. In the second step, the transmitted codeword is formed by geometric construction. Thus, the transmitted codeword is a one-dimensional hologram of a point.

The formed digital hologram is transmitted over the communication channel with interference and on the receiving side decoding is performed - hologram restoration, search for the maximum and output of its coordinate in the form of n-bit output code.

6.2 Modeling results

The study of noise immunity of the considered holographic code is carried out by modeling in Matlab environment the processes of coding-decoding during the transmission of code messages through a channel with noise. A pseudorandom noise generator realized by Random function is used as a noise source.

Figure 8.4 shows the result of recovering the signal carrying the value Y=100. There is a small coding noise in the reconstructed signal.

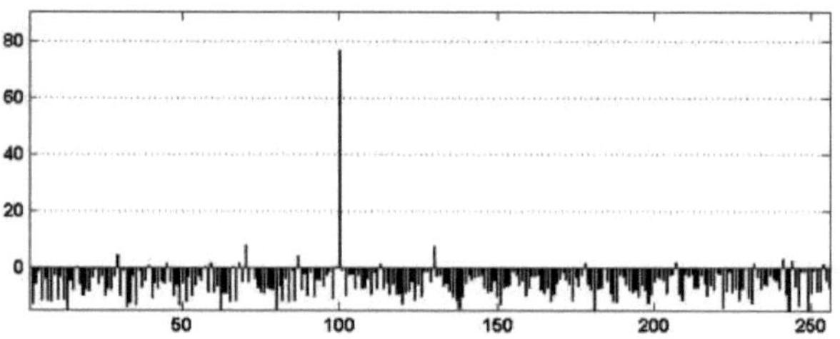

Figure 8.4. Reconstructed signal at N=256

The level of coding noise, which determines the potential noise immunity of the code, depends on the length of the code combination. The shape of the reconstructed signal carrying the value Y=400, at N=1024 (redundancy factor q=128) is shown in Figure 8.5.

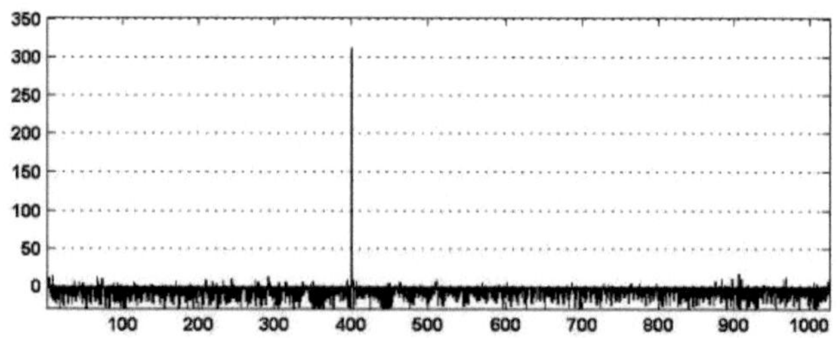

Figure 8.5. Reconstructed signal at N=1024

Let us consider the stability of the code to errors caused by noise in the communication channel. The impact of noise is modeled by replacing a part of the hologram with a binary random sequence. When the length of the random fragment in the signal is 70%, the recovered signal exceeds the maximum value of the noise emission by 11 dB (Figure 8.6).

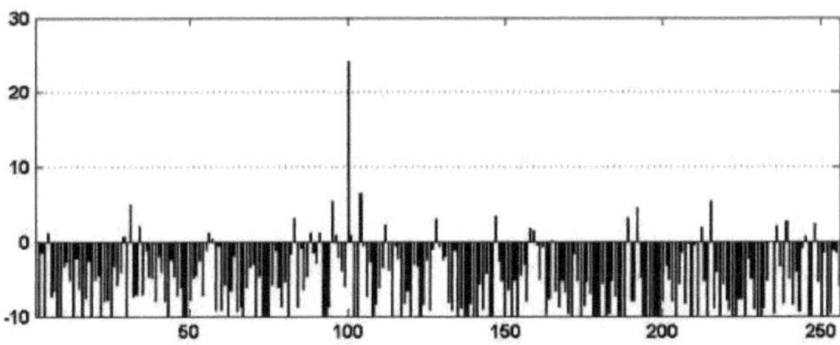

Figure 8.6. Reconstructed signal at random fragment length of 70%

Increasing the length of the code combination leads to an increase in the noise immunity of the code. At N=1024 the output signal is reconstructed from the signal, on 80% consisting of a random sequence. Signal-to-noise peak ratio at the decoder output is equal to 6.3 dB (Figure 8.7).

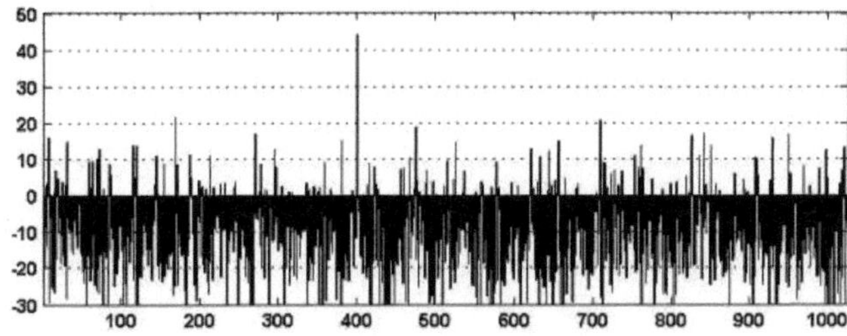

Fig. 8.7. Signal at the decoder output at N=1024. Length of random fragment 80%

If the noise in the channel distorts 90% of the signal, it is necessary to increase the length of the code combination to N=4096 to obtain a signal-to-noise ratio at the decoder output of 5 dB (Figure 8.8).

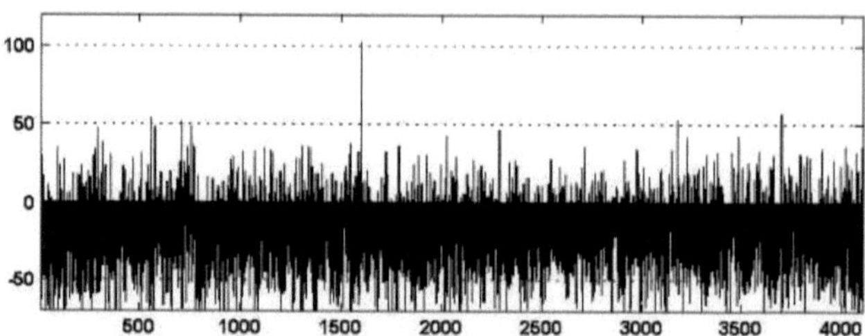

Figure 8.8. Losses in the 90% channel. Reconstructed signal at N=4096

Holographic coding provides the possibility of information transmission over noisy communication channels with a large number of errors. The maximum allowable noise level is determined by the amount of redundancy introduced during coding - the length of the code combination. Boundary possibilities of known systematic binary linear linear block codes are defined by Singleton's theorem, according to which for error correction the code must have at least two check symbols per error. This means that the number of errors in the message must be less than 50% to maintain error correction capability. The capabilities of the proposed holographic code are determined by the level of redundancy introduced: at the redundancy factor $q = 32$ error-free decoding of the signal occurs when replacing 70% of the signal with a random sequence, while the Hamming code used in GLONASS corrects one error in an 85-bit word.

The conducted researches have shown that the introduction of holographic coding in the satellite communication channel of GLONASS system will give an opportunity to navigation equipment of consumers to receive information from more satellites due to the ability of the code to correctly decode the signal at replacement by a random sequence up to 70% of the signal duration, which will significantly increase the accuracy of positioning. To realize this method it is necessary to introduce additional noise-resistant holographic coding into the communication channel and to modify the software of navigation equipment of consumers by installing a module of holographic decoding.

7 Using holographic coding to improve data rates in mobile networks

Nowadays, the main way of data transmission is through mobile communication networks. In analog mobile networks of the first generation, the communication range was determined primarily by the sensitivity of the receiver of the mobile terminal and the noise level in the communication channel, more precisely by the signal-to-noise ratio. In digital mobile networks (2G generation and beyond) these characteristics still remain basic and are used in the development of standards and communication equipment in the choice of modulation types, spectral efficiency, bit rate of data transmission, the choice of methods of noise-resistant coding for error detection and correction. In second- and third-generation networks, the bulk of traffic was voice transmission. Errors occurring in the digital communication channel led to deterioration of communication quality, but had no effect on communication speed. The communication range (the size of the network coverage area) was determined by the boundary of acceptability of voice transmission quality.

To date, the development of radio communications has come to the following:

- all communication systems are digital

- practically all communication systems are included in the Internet and therefore at the fourth (transport) layer of the Open Systems Interconnection model (OSI model) use the TCP protocol.

- The theory of electrical communication considers the processes of signal transmission at the first (physical) layer of OSI model (modulation, spectral efficiency, bit rate) and at the second (channel) layer (detection and correction of errors occurring at the physical layer). The ultimate goal of these layers is to provide the required transmission rate of multi-bit data blocks with the required bit error probability

- starting from the third layer of OSI model and above - this is the IT sphere. At the fourth layer, the TCP protocol provides guaranteed delivery of IP packets

- the choice of solutions for the construction of the communication channel (type of modulation, method of noise-resistant coding, coding speed, redundancy level) is not coordinated with the parameters of the TCP protocol (packet size, waiting time for a receipt, redundancy level arising due to duplication of packets) and, as a result, with the resulting speed of information transmission at the application layer.

In 4G-5G networks, the bulk of traffic is image transmission and voice is also transmitted as packet data in IP networks. Modulation methods, processing methods and noise immunity coding have changed. However, the block size used for data transmission has changed little - it still ranges from a few bits to a few tens of bits. To blocks of this size is applied noise-tolerant coding and the probability of residual error after decoding determines the quality of the communication channel. Usually error probability less than 10^{-4} is considered acceptable and a channel with error probability 10^{-5} is considered good. This estimation is made from the point of view of the theory of electrical communication for the methods of multiplexing, modulation, noise-tolerant coding applied in the radio channel. The error probability of 10^{-4} approximately corresponds to the boundary of the 4G network coverage area. At the same time, the data transmission speed is hundreds of times lower than the 300 Mbit/s speed declared by the standard and 4G operators and may well be less than 1 Mbit/s. Hereinafter we are talking about the data transmission speed determined by the channel quality, not the number of active subscribers, and it should be measured during the hours of lowest load.

7.1 Problem statement

This drop in speed is explained by the different approach to the quality of data transmission in communication networks and in IP-networks - if in communication networks a very small but non-zero probability of error is acceptable, then in IP-networks TCP protocol provides guaranteed delivery, and the data packet is delivered either without errors or a communication break is fixed. In the TCP protocol, a data packet is 11680 bits in size. This means that with a probability of error in a digital communication channel 10^{-4} each transmitted packet will arrive with an error in at least one bit. Single errors are eliminated by the correction codes used, but double errors or more are transmitted by the TCP protocol, which provides guaranteed delivery by duplicating packets. A similar number of errors occur during retransmission. Due to the random nature of the errors, individual packets will be received correctly and will pass further into the network, but this results in tens or hundreds of times slower information transfer rates to the user (Figure 9.1). In 5G and 6G networks, there is a tendency for packet size to increase, so the same number of errors will result in an even greater reduction in speed.

When designing existing and future communication networks (5G and 6G), technical solutions for achieving the required communication speeds (gigabit speeds in 5G networks and terabit speeds in 6G networks) are incorporated into them. A large number of articles are devoted to the comparative analysis of noise immunity of used and prospective radio communication technologies, but in most of them the probability of errors occurring in the communication channel (bit error rate) is considered in the range of 10^{-4} and more (Fig. 9.2 [136]).

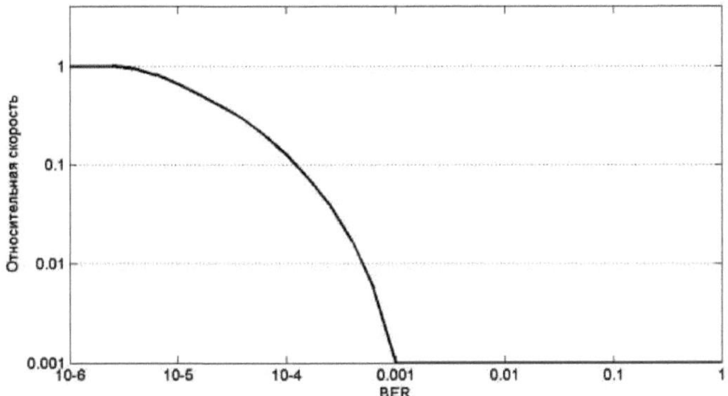

Figure 9.1. Dependence of relative data transmission rate on error probability at packet size at the transport layer

11680 bits

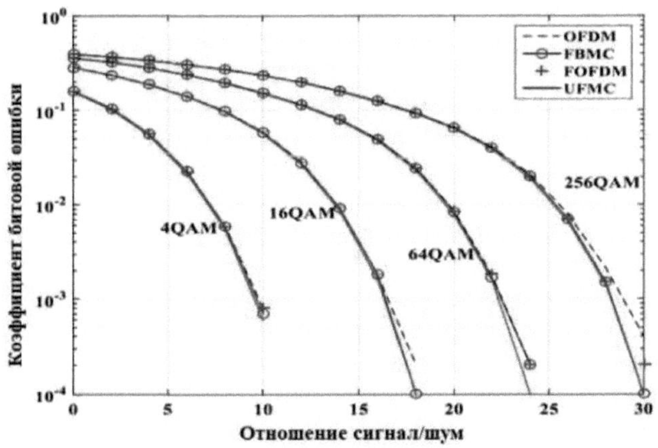

Figure 9.2. Efficiency of multi-station access options for mobile networks

In mobile data networks of all operators, there are many parts of the service area where the actual data rate is units of Mbps, i.e. 100 times lower than the speed defined by the 4G standard. This leads to the following:

- communication channels are busy with retransmission of faulty packets, network capacity is used only by 1%.

- subscribers in better conditions in terms of communication quality, but using the same channel, for example, in the same building, but on a higher floor, receive lower speed due to the fact that the communication channel is occupied by retransmission of failed packets.

- Due to the lack of channel capacity and subscribers' lack of promised speed, operators are forced to look for additional frequency resources and build additional base stations.

7.2 Decision

The decision of this problem is increase of noise immunity of a communication channel with decrease of probability of an error at output of the decoder to 10^{-6} . For such essential increase of noise immunity in conditions of a weak signal and low signal-to-noise ratio the method of noise-immune coding with much more correcting ability in comparison with known, applied in communication networks, codes is necessary.

Passport data transmission rate in 5G networks is provided at good channel quality (at signal-to-noise ratio S/N=20dB error probability 0.01 for 64QAM) within a radius of 100 m from the base station. When the distance is increased by 3 times the field strength decreases by 10 dB, in this case S/N=10 dB, the error probability is 0.12-0.15. Interference-resistant codes used at the channel level, with redundancy, as a rule, not more than 2, do not cope well with this level of errors. Widespread codes from Reed-Solomon code to LDPC produce a large number of decoding errors in such a channel, which are corrected by TCP protocol by repeatedly duplicating IP-packets. If in LTE network instead of 300 Mbit/s a subscriber receives real speed of 3 Mbit/s, it means that on average each IP-packet is repeated 100 times, i.e. 100-fold redundancy is introduced.

With the claimed range of the 5G base station of 1 km, the zone of high-speed communication with a radius of 300 m has an area of 9% of the coverage area of the base station. Thus, the real data transmission speed on 90% of the territory of the mobile network corresponds to the speed of the previous generation (for LTE - 3G speed, for 5G - LTE speed), i.e. there is a territorial digital inequality.

To solve this problem it is necessary when choosing network technologies to consider the impact on the speed of data transmission laid signal processing methods at all layers of OSI model, providing information transfer - from the first (physical) to the fourth (transport).

Such a method is the holographic coding method [44]. The holographic coding method is based on mathematical modeling of a digital one-dimensional hologram created in virtual space by a wave from an object whose image represents the input data block. The hologram is formed by constructing a digital sequence corresponding to a Fresnel zone ruler. Thus, a one-dimensional object A(i) is matched with a one-dimensional hologram H(j). The values of the calculated hologram are rounded to one bit - positive ones are taken as 1, negative ones - as 0. As a result, an n-bit one-dimensional array H_O (j) is formed, which is a code combination corresponding to the k-bit input data block X

The peculiarity of the holographic code is that the input coded word must be represented in a single position code. In this case, the optical object for which the hologram is constructed is a point source on a black background, and the information is embedded in the coordinates of a point on the object field. The result of coding is the simplest hologram - a Fresnel zone plate, the coordinates of the center of which carry the encoded information. In the process of encoding, the symbol interval (used for transmission of a symbol, data block) is divided into L slots of duration t. The digital input data block X to be transmitted, which is a k-bit binary code, is converted into a position code A, consisting of $n = 2^k$ points A(i), i = 1,..., n, the value of one of which is 1, the others are zeros: A(i) = 1 at i = X, A(i)=0 at i ≠ X. As a result, block A has (n-1) zeros and one unit at the position given by block X. Thus, the input data block is used as the address of the unit position in the sequence of zeros of the unit position code.

Decoding in the receiver takes place as follows. The signal received during the symbol interval is subjected to the digital hologram reconstruction procedure. The resulting digital line array has a maximum in cell Y with a number corresponding to the value of the input block X. From an optical point of view, the reconstructed object is a dark line with one bright point and a small background illumination at the other points. The holographic code utilizes the hologram divisibility property and is able to recover the transmitted image from the hologram fragment as well as the one hidden by noise. Restoration capabilities of the holographic code depend on the hologram length (the number of slots L in the symbol interval).

In digital communication channels often more informative is an estimation of noise immunity not on a signal/noise ratio, and on a limit quantity of random errors in the decoded word. As a result of modeling it is established, that decoding of a codeword of length n=256 occurs error-free at distortion of 80 bits (31%).

When using holographic coding with 10-fold redundancy, the probability of error at the decoder output is 10^{-6} with the probability of error in the channel 0.2. As a result, in a communication channel with a data transmission rate of 3 Mbit/s (reduction of the rate 100 times from the maximum rate) at inclusion of the holographic code the rate will be 30 Mbit/s (Fig. 9.3). At the same time:

- the average speed will increase by 10 times (it will be less than the maximum speed not by 100, but by 10 times)

- channel utilization will drop by a factor of 10

- freeing the channels will allow to do without additional frequency resource and increasing the number of base stations.

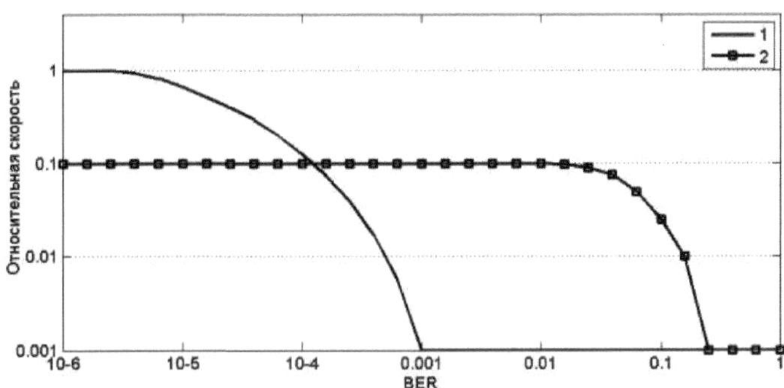

Figure 9.3. Relative data transfer rate:

1 - without additional holographic coding,

2 - with holographic coding

In order that the redundancy of the holographic code does not reduce the data transmission rate in channels with good communication quality, it should be included only in those channels where the rate drop is recorded more than 10 times. In this case, the dependence of the relative data transmission rate on the bit error rate looks as shown in Fig. 9.4.

Implementation of the method consists in modifying the software of network equipment and mobile terminals.

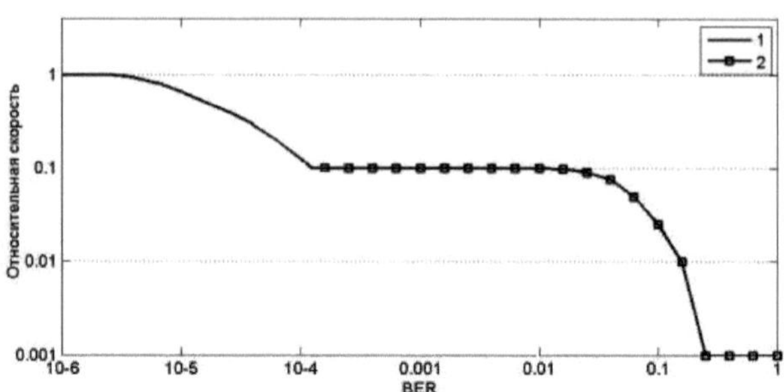

Figure 9.4. Relative data transfer rate:

1 - without additional holographic coding,

2 - with holographic coding

8 Holographic methods of parallel information transmission

A common way of transmitting information by means of electromagnetic radiation in both optical and radio bands, through waveguides and in open space is serial transmission. Frequency division of channels in the radio band, as well as WDM technology in the optical band, does not form a parallel transmission in one channel, but creates a number of serial channels.

Practically the only way of parallel transmission of information is the transmission of a two-dimensional image through a waveguide, which can be of different designs - from metallic to fiber-optic [77]. Optical multimode waveguides have the ability to reproduce the image of an object located in its input section ($z = 0$) in a sequence of in-phase sections distant from the input by distances z_s - sL (L is the distance to the first in-phase section depending on the type of waveguide and its parameters, s is the serial number of the in-phase section). In recent years, there are more and more publications that consider the possibility of image transmission over multimode fiber [137-140]. For example, a typical fiber with a core diameter of 100 μm can carry up to 10,000 modes and in principle transmit an image with approximately the same number of pixels [85]. However, in such a fiber, each of the individual modes propagates at a slightly different rate, resulting in amplitude and phase distortions of the image and the formation of a speckle structure. Full a priori knowledge of the input image and fiber details may allow us to numerically simulate optical propagation, reconstruct the transmission matrix, and then decode the output data, but in practice this is not feasible due to the need for very large computational resources.

At the same time, in optics and other fields using wave processes, there exists and uses an effect that can be considered as a parallel transmission of information - holography. The uniqueness of holography consists in the fact that information about the initial object is transmitted in space by a monochromatic wavefront (i.e., in one frequency channel) and forms a large volume interference pattern to be registered for a time equal to one period of the reference wave (in optical holography - for 0.002 picoseconds).

8.1   Problem statement

Let us consider a task more general than image transmission - transmission of arbitrary digital information. In existing systems for transmission over communication channels, the initial digital information block is represented as a one-dimensional array. In the process of transmission, the information block enters the communication channel element by element and thus unfolds in time, turning into a signal, the duration of which is proportional to the number of elements of the array. The equipment for transmission and reception is localized in two points of space (light source and photodetector in the optical channel, transmitter and receiver antenna in the radio channel). The process of hologram formation is different from the process of information transmission. Nevertheless, holography can be considered as a method of information transfer from the region of space where the object is located to the region of space where its hologram is formed. In hologram formation, the transmitted information block is a 2- or 3-dimensional image of the object, which is transmitted in one spatial channel and forms the hologram matrix. In classical holography, information is transmitted in space in parallel (all points simultaneously) when a hologram is created. But this process can be distributed in time by transmitting information about the object through communication channels to synthesize the hologram at the receiving end, e.g., in [140], depth maps and surface textures of the recorded object are transmitted through the communication channel.

If holography is used for information transmission, then images of objects on the transmitting side are formed one after another with some clock frequency and holograms are registered on the receiving side with the same frequency. It is possible to consider that at holographic information transmission there is a transformation of space-time matrix which gives a possibility of parallel transmission (information sweep in time, used at sequential transmission, is replaced by sweep in space). Thus, the information is deployed in space by the area of the received hologram, and in time two local points are used - the moment of object formation and the moment of hologram formation.

Holography provides a choice of two options for the location of digital holographic processing. The first option is direct holographic transformation - formation of an image in the transmitter plane, propagation of the wavefront, registration of the interference pattern in the receiver and restoration of the original image by digital hologram processing. The second variant - inverse holographic transformation - formation in the transmitter plane of the hologram of the original image and transmission of the hologram wavefront. In this case the original image is formed in the receiver plane. The choice of the variant depends on where there are more available computational resources - in the receiver or in the transmitter.

Limitations. The first is the lack of electronic means capable of transmitting and processing information at the speed of hologram transmission in open space. The second - registration of holograms of real images requires multielement photomatrices with a large dynamic range of brightness. Therefore, at the first stage of research of this technology the rational solution is to reduce the volume of transmitted information and use the simplest images - single-bit matrices of the initial object, for example, 16x16, for coding messages. When using a position code, as, for example, in holographic coding [44], the hologram of the simplest object (one luminous point) is an image of a Fresnel zone plate (Fig. 10.1).

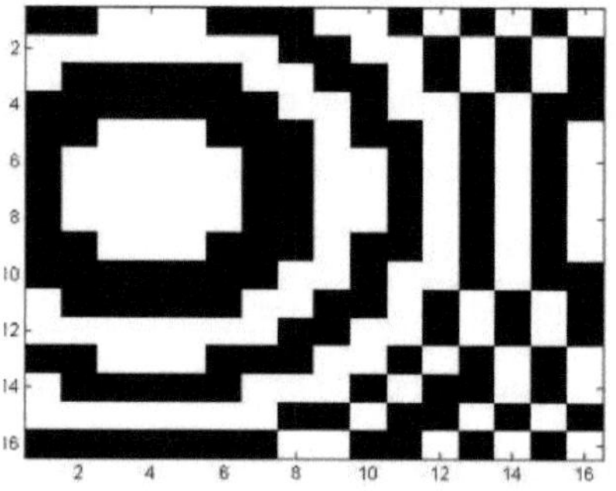

Fig. 10.1. Fresnel zone plate

In this case the useful volume of transmitted information is 8 bits (two 4-bit coordinates of the luminous point on the object image of 16x16 size, corresponding to the coordinates of the center of Fresnel zones on the hologram).

In the receiver plane the coordinates of the center of Fresnel zones are calculated by digital processing, in this case it is the hologram reconstruction. If hardware resources allow using a matrix of light sources, a hologram is formed in the transmitter plane, and the interference pattern in the receiver plane is an image of a point source, the coordinates of which are the transmitted information.

8.2   Variants of solving the problem of parallel information transmission

1- Optical holography. For transmission in open space, a single-bit binary optical matrix of the object is formed in the transmitter by the input block of the transmitted information. Matrix elements in the state "1" are coherent light sources formed by optical modulators of the plane front created by the laser, in the state "0" there is no radiation. The wave front formed by spherical waves of the object matrix elements forms a hologram in the receiver plane, which is fixed by the photodetector matrix. The speed of information transmission is determined by the speed of optical modulators and photodetectors. In [141] a spatial light modulator is described, which focuses the beam in a given direction and changes the light intensity several orders of magnitude faster than commercial technologies on liquid crystals or micro-mirrors. The transmission range is determined by the ratio between the resolution of the hologram recording medium and the wavelength of the radiation used - in optical holography the size of the photodetector matrix, the number and sensitivity of its elements and is several meters. It is possible to increase the range many times by using narrowly directed lasers as elements of the transmitter matrix. An array of lasers pointed at the receiver matrix forms an interference pattern in its plane. The range in this case will be determined by atmospheric turbulences and correspond to the range of atmospheric optical communication lines. The noise immunity of a serial optical channel with holographic coding is considered in [102].

If success is achieved in the field of image transmission over multimode fiber and these researches get practical realization, holographic methods of multimode parallel transmission of arbitrary digital information will also be developed. Unlike real images requiring high spatial resolution, digital information in multimode mode can be transmitted in the form of small holograms ranging in size from 8x8 to 32x32. The problem of mode dispersion and other types of distortions arising in fiber transmission is solved in this case due to the high distortion resistance of the hologram. Fig. 10.2 shows a hologram of an object having 4 luminous points.

Fig. 10.2: Hologram of four point objects

The size of the hologram is 32x32, so the coordinates of each point carry 10 bits of information, respectively, 4 points provide the transmission of 40 bits. Fig. 10.3 shows the result of this hologram reconstruction obtained by modeling the process of parallel information transmission in MATLAB environment.

Thus, the use of optical holography makes it possible to organize parallel transmission of information both in open space and in multimode fiber. High noise immunity of holographic transmission allows to resist atmospheric interference and mode dispersion and to increase by 40 and more times the speed of information transmission in comparison with serial transmission in the same propagation medium [142].

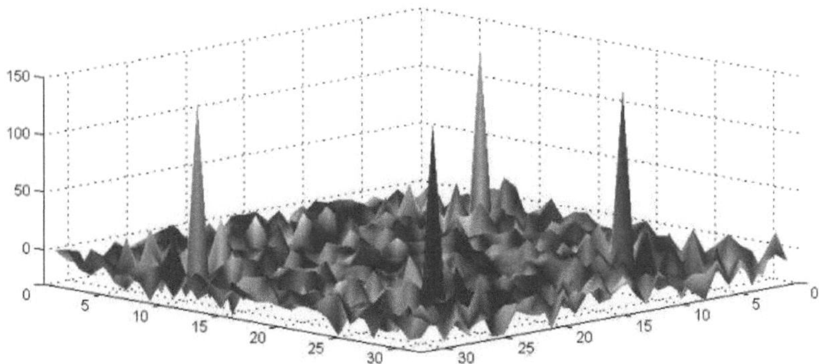

Fig. 10.3. Image of four point objects reconstructed from the hologram

2. Radio holography. The initial stage of development of parallel transmission of information over a radio channel is the technology of space-time coding (STC). Space-time coding is realized in systems with several antennas on the transmitting side and several antennas on the receiving side, in the so-called MIMO (Multiple Input - Multiple Output) systems. In [143], the possibility of reducing the size of MIMO antenna devices by using Holographic MIMO Surfaces (HMIMOS) was considered. Their development did not aim to create a channel for parallel transmission of information, but they may well find such an application.

Formation and registration of the wavefront occurs in the radio band. The transmitter and receiver matrices are antenna arrays or HMIMOS surfaces. The size of the antenna devices is determined by the operating frequency range. With the terahertz band planned for use in 6G networks, the size of the antennas will be 10-30 centimeters, which is acceptable for many fixed communication nodes. The antenna in this case

101

is not a phased antenna array, its elements operate with a constant phase, which is measured at the receiving end during the joint alignment of the transmitting and receiving antenna system. Each element of the antenna array fixes the intensity of the interference field in one point, the whole array forms the received hologram. As a result of subsequent digital processing the transmitted information block is restored.

3. **frequency holography**. Another variant of transformation of the space-time matrix is the transfer of information interaction from the space-time domain to the time-frequency domain. Instead of the considered transfer of the information block from a linear array in time to the spatial matrix of the hologram, we can use the transformation of the information block into a linear array of spectrum in the frequency domain. This is somewhat associated with wavelength-division multiplexing (WDM) in optics, frequency-division multiplexing (FDM) in radio channels, and broadband technologies such as code division, linear frequency modulation, and others, but while retaining the holographic approach corresponds to spectral holographic coding. In this case, the hologram is built in frequency space - the shape of the hologram corresponding to the transmitted information block has a signal spectrum. On the transmitting side the initial information block, for example, a byte, is translated into a single position code, which is a 256-bit code containing 255 zeros and one unit, the position number of which is given by the initial byte. On this one-dimensional array a one-dimensional (linear) hologram is constructed, the values of which are rounded to one bit, i.e. the hologram is a 256-bit sequence of zeros and ones contained in approximately equal proportion. To transmit the hologram through the communication channel it is necessary to generate a signal y(mT), the shape of the spectrum of which is the same digital one-dimensional hologram - one in the i-th position of

the hologram means the presence in the spectrum of the i-th harmonic, zero - the absence. This function is realized by adding harmonics with corresponding numbers:

$$y(mT) = \sum_{i=1}^{N}(G(i) \cdot \sin(2\pi(m / M) \cdot i + r(i) \cdot 2\pi)),$$

where G(i) is the linear array of the hologram, N is the number of harmonics, M is the number of samples in the signal, r(i) is a random number in the range 0...1.

This method of signal conditioning is a type of orthogonal frequency division multiplexing (OFDM), characterized by the fact that the frequencies of the N orthogonal subcarriers are in multiples, and amplitude manipulation is used as digital modulation.

When using the signal y(mT), the duration of which is equal to the integer number of periods of all used harmonics, its spectrum has a linear form corresponding to the hologram of the virtual image of the input data block (Fig. 10.4).

Fig. 10.4. Spectrum in the form of a hologram

The signal y(mT) can be synthesized in analog form, but in many cases it is more accurate and easier to perform digital synthesis, which can be implemented in two ways. The first method is the algebraic addition of N carriers forming a linear spectrum corresponding to the hologram, and the second method is the inverse fast Fourier transform of the hologram.

An important signal parameter is the peak factor, which reaches its maximum at zero initial phase of the used carriers. A higher value of the peak factor places higher demands on the linearity of the amplifier. The best result is achieved when the initial phases of the carriers are randomly distributed. The signal in this case has a noise-like shape (Fig. 10.5).

In the receiver the spectrum of the received signal is calculated. The digital array representing the spectrum is considered as a one-dimensional hologram of the original digital block, and decoded by the holographic method - restoration of the original data block by digital hologram is performed.

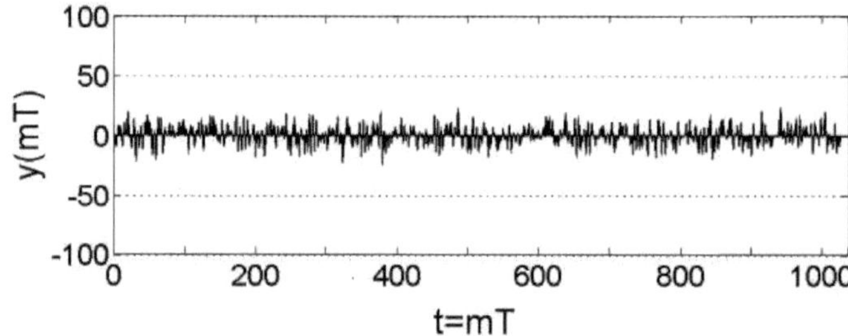

Fig. 10.5. Signal waveform with random initial phase of carriers

Limitations of this method: the need for a wide range of frequencies and the existence of a limitation in the speed of information transmission due to the requirement for a signal duration of at least the whole period of the lower harmonic of the frequency range used.

The proposed methods of transition from serial to parallel transmission of information in optical and radio bands using holographic transformation of arbitrary digital information allow to significantly increase the speed of data transmission. The theoretical limit of the throughput capacity of such communication channels is the speed of transmission of a hologram - the amount of information contained in a three-dimensional image of a complex object, during the duration of one period of a wave of electromagnetic radiation. The achievable gain in speed depends on the size of the hologram, the frequency range used and the selected coding parameters. The development of hardware solutions for the holographic method of information transmission is the subject of further research.

9 Optical radio

In almost all communication systems, there are tasks that cannot be solved due to the limited speed of information transmission. With the development of communication there is an increase in the available transmission speed, some tasks are solved, but their place is taken by others. This state of affairs is considered to be natural, and all efforts of developers are aimed at increasing the bandwidth of communication channels. At the same time there are processes with incomparably higher speed of information transmission, but their transmission principles are not used in technical systems. For example, when photographing the starry sky, information about the whole part of the universe in the field of view is transmitted to the registering device at the speed of light. The fact that only a few tens or hundreds of megabytes of this practically infinite amount of information can be recorded speaks about the capabilities of the recording device used, not about the speed of information transmission in space.

Transmission of information by means of electromagnetic radiation occurs in two main ways - optical (in the visible part of the spectrum with the inclusion of neighboring areas - UV and IR) and radio (in the long-wave part of the spectrum). Let us compare the distinctive features of these methods and consider the possibility of using the third, unifying method.

9.1 Optical transmission of information

It is the basis of vision of living beings and the whole technique of fixing images, starting with photography. Although usually vision and photography are not referred to the process of information transfer, in fact it is the main way of information transfer - from the area of space, in which the observed object is located, to the area of perception - on the retina of the eye or photographic matrix of the camera.

An illumination source is required to form an image. Each point on the surface of each object in the optical scene emits, in general, a spherical wave, reflecting the radiation of the source. Thus, the entire space of the optical scene is filled with radiation of the same spectrum (the spectrum of the illumination source minus spectral absorption by the objects) propagating in all directions. During propagation, the intersecting waves do not interact due to the linearity of Maxwell's equations, but interference occurs on any surface (screen) where they hit. With a monochrome source the interference pattern is clearly visible (this is used in holography), with a non-monochrome source the screen is uniformly illuminated (interference is indistinguishable due to the addition of incoherent waves). To obtain an image of an optical scene, a lens (in the simplest case, a lens) is needed. The lens realizes spatial separation of waves falling on it in such a way that in each point of the screen (the surface on which the received image is formed) comes a wave from only one point of the optical scene (one spatial direction). Therefore, there is no interference on the screen. In monochrome illumination a monochrome image is formed, in non-monochrome illumination a multicolor image is formed.

Features of the optical method:

- There is no modulation, multiplexing, frequency division or any transformation of the carrier (the wave emitted by the illumination source). We can speak only of amplitude modulation, since the brightness of each point of the image is proportional to the level of incoming radiation

- The absence of signal conversion operations gives zero modulation/demodulation time and theoretically unlimited speed of image transmission. The entire image of any great complexity is formed on the screen in one period of the light wave. The speed of light remains a limiting factor, but given the amount of information contained in the image and transmitted in parallel simultaneously, the bandwidth of the channel in the absence of noise is virtually unlimited. The special theory of relativity limits the physical speed of signal transmission for clock registration between two spatially distant observers [144], but does not prohibit the transmission of information with any large bit rate.

9.2 Serial transmission via radio channel

The development of radio since its invention went as an alternative to wire communication (replacing the medium of transmission - wire to ether) with preservation of the principle of information transmission from one point of space to another (radio telegraph), but not as a choice of an alternative part of the electromagnetic spectrum with preservation of the optical method of transmission from the transmitter's space to the receiver's plane, when the whole message is transmitted in parallel simultaneously. Therefore, the information transmitted over a radio channel must be represented in serial form for transmission over a serial channel. If only one radio channel is used in the considered region of space, the main problem is to increase the speed of information transmission, for which a large number of modulation methods have been developed. If in this region of space it is necessary to organize several communication channels, the problem of their interference in the receiving antenna arises. The main way to solve this problem is frequency separation, the possibilities of which are always limited by the width of the allocated frequency range. To further increase the number of channels in conditions of frequency resource deficit, time, phase and code separation of channels is used.

9.3 Optical radio

Radio vision, or introscopy [145-147], is concerned with obtaining visible images of objects using radio waves to study the internal structure of objects that are opaque in the optical wave range and to observe objects in an optically opaque medium. Various physical effects and phenomena are used in radio vision, and liquid crystals, semiconductor single crystals, special photographic films, etc. are used as a sensitive element. All such elements under the influence of radio waves change their optical characteristics - reflection coefficient or transparency for visible light. Most often radio images of objects are obtained by scanning a narrow beam of radio waves and receiving signals reflected from the object. However, in radio vision and introscopy the task of transmitting arbitrary digital information is not set.

Recently, more and more attention has been paid to the systems for surveillance of the surrounding space using radio-band radiation in the tasks of space navigation of robots for various purposes, mobile unmanned vehicles, control over the movement of objects in a limited space, etc. As a rule, active systems are used for these purposes, which include a transmitter, illuminating object and receiver (or, alternatively, a dispersed receiver system) [148]. In this case, the term "radio light" is used, which is understood as a local artificially created field of broadband incoherent in space and time radiation in the radio wavelength range. Hitting nearby surfaces and objects, the radiation is partially absorbed and passes through them, as well as partially reflected. As it propagates further, it carries information about the medium with which it interacts. In this case, the situation is similar to that of ordinary (visible) light, the only difference being that it is a different frequency range [149].

The possibility of creating an analog of a biological eye in the radio band for observing the surrounding space in artificial radio light is considered in [148]. A number of studies have shown the possibility of observing objects in radio light using multibeam systems similar in their properties to the eyes of biological objects with a small number of sensitive elements. A common approach for works in this area is the use of antenna systems with a narrow radiation pattern [150-152]. In [148], an experimental model of a multibeam radio-light imaging device based on a Rothman lens is proposed, which allows simultaneous acquisition of information from each beam. The disadvantages of this approach include the lack of scalability of the imaging system (both the number of beams and their directionality are rigidly set by the geometry of the system, which cannot be flexibly changed) and the complex design of the lens in the case of the need to work with two-dimensional images. To expand the possibilities of imaging methods in radio light, it is necessary to be able to scalable formation of multiple beams simultaneously with high resolution. The basis for a receiving system with such characteristics can be based on radio astronomical methods, in which a relatively small number of receiving antennas form images of space with high resolution [153]. In this case, radiolenses are of increasing interest. Lens antennas, and primarily Luneberg lens antennas [154], are a promising type of antennas for 5G and 6G networks [155]. In addition, Luneberg lenses are used as pattern forming devices in radar [156], replacing phased antenna arrays, and radio telescopes.

The Luneberg lens is an inhomogeneous dielectric sphere in which the refractive index n varies according to the law

$$n(r) = \sqrt{\varepsilon(r)} = \sqrt{2 - \left(\frac{r}{a}\right)^2} = \sqrt{2 - (\bar{r})^2},$$

where $\varepsilon(r)$ - is the relative dielectric constant of the lens material at the point r, $\bar{r}$ - is the current radial coordinate, a is the radius of the sphere. The main property of the Luneberg lens is that when a beam of parallel rays falls on it, all of them will gather in one point - the focus.

The shape of a radio lens is determined by the required law of change of refractive index n (the ratio of phase velocities of radio wave propagation in vacuum and lens). When n > 1, the radiolens (as well as the lens in optics) is called retarding, and when n < 1 - accelerating (the latter has no analogues in optics). Decelerating radiolenses are made of homogeneous dielectric materials with low losses (polystyrene, fluoroplastic, etc.). Accelerating radiolenses are made of metal plates of a certain shape. Their principle of operation is explained by the dependence of the phase velocity of an electromagnetic wave propagating between parallel metal plates on the distance between them, if the vector of its electric field is parallel to the plates. The lens body may consist of discrete elements of cubic shape with different refractive indices. Thus, the required law of refractive index variation can be realized with a sufficient degree of accuracy in the form of a radially variable cluster of inhomogeneities in the dielectric. Examples of lenses are shown in Figure 11.1 [157].

Fig. 11.1 Variants of Luneberg lens execution

However, most of the works consider and use only one application of radiolenses - formation of one or several beams with a narrow directivity diagram, while the optical method of information transmission in other areas of the electromagnetic spectrum remains unexplored. Such a way is the use of the laws of geometrical optics in the radio band. This method can be defined by the term "optical radio", which was first used by Nobel laureate N.G. Basov in 1961 [158]. To date, the development of radio electronics in the high-frequency region of the electromagnetic spectrum has reached a level where it is possible to begin the practical development of optical radio.

In the transition from optics in the visible part of the electromagnetic spectrum to optical radio in the radio band, only the size and material of the lens and the size of the receiver (receiving matrix) change.

A transmitter in optical radio is an array of N transmitters emitting signals at the same frequencies. Any modulation and multiplexing techniques can be used in each channel. The antennas can also be of any type, but if directional antennas are used, their patterns must match in direction.

A receiver is an array of N receivers working in pair with its transmitter and forming N communication channels. The main element of the array is a radio lens that forms a radio image of the transmitter antenna array in the plane of the receiver antenna array. In accordance with the laws of geometrical optics, if the wavelength ratio and lens size are correctly selected and the system is correctly aligned, only one transmitter's radiation enters each receiver antenna.

Features of optical radio:

- In one frequency channel in one area of space, it is possible to organize many channels of information transmission on the same carrier that do not interfere with each other

- The sensitivity of the receiver is determined by the area of the lens, and to convert the radiation energy collected by the lens into an electrical signal in the input circuit of the receiver, an elementary antenna - a cell of the receiving matrix - is sufficient for this purpose

- The antenna matrix of transmitter antennas may consist of directional antennas with high gain, the antenna matrix of receiver antennas may consist of elementary non-directional antennas

- Main type of communication - within line of sight (periscope principle may be used)

- System alignment is necessary to ensure that the antenna of each receiver receives the signal of the correct transmitter

Possible applications:

- Multiple radio relay lines (RRLs) operating on a single frequency. An example of possible realization on serial equipment of 38 GHz RRL: 9 transmitting antennas 30 cm in diameter are assembled in a 3x3 array like floodlights at a stadium. Receiver antennas are similarly assembled. In front of the receiver array is installed a radio lens with a diameter of about 1 meter, so the radiation of each transmitter antenna hits only one receiver antenna, and there is no interference. As a result, the 9 trunks operate on the same frequency

- Communication channels between fixed nodes (base stations, data kiosks in 6G networks)
- Radio telescopes
- Radars

The main element of optical radio is a radiolens, and unlike the aperture antenna of traditional radio, it is used to form a radio image projected onto an array of elementary antennas. This makes it possible to organize in one spatial channel a large number of radio channels operating on a single carrier with an independent choice of modulation methods and the width of the used spectrum. A limitation of optical radio is the need for line of sight, but this is a conditional disadvantage. On the one hand, in widely used radio relay lines, for example, not only direct visibility is required, but also very precise (to fractions of a degree) antenna alignment. On the other hand, humans receive the bulk of information through vision and do not consider it a disadvantage to have to look in the direction from which they receive information. Therefore, it can be assumed that with the development of high-speed directional methods of information transmission in 6G networks (optical radio), the need to maintain line of sight will no longer be perceived as a disadvantage.

10 Realization of stable absolute phase modulation by means of holographic coding

From the theory of communication it is known that phase manipulation (FMn) is characterized by high noise immunity. In 1946, V. A. Kotelnikov in his doctoral dissertation "Theory of potential noise immunity" proved that the FMn signal with 180° manipulation is the best way to transmit binary signals and achieves potential noise immunity [159]. However, the realization of a demodulator for coherent reception of such a signal is complicated by the need to maintain the equality of phases of the reference oscillator and the incoming signal. In practical schemes the reference signal is formed from the received oscillation. In this case, all schemes for the formation of the reference signal are such that due to various uncontrollable factors are possible random changes in the sign of the reference signal. This means that the characters recorded at the receiver output, even in the absence of additive interference in the channel after a random jump in the phase of the reference signal inverted. This will continue until the next phase jump of the reference signal. The so-called "inverse operation" phenomenon occurs, which severely limits the application of absolute FMn (AFMn) in communication systems. Therefore, the 180° AFMn, although it provides the highest possible noise immunity of radio communications, is not used in practice because of the "reverse operation" of the coherent detector [160].

In 1954, N.T. Petrovich proposed a method of relative phase manipulation (RPMn), which eliminates the phenomenon of "reverse work" [161]. This method is widely used in modern digital navigation, communication and television systems. The use of OFMn signals is incorporated in communication standards DVB-S, DVB-S2/S2X,

GLONASS, CDMA, Wi-Fi IEEE 802.11 and others. [162]. In all these cases, the task of message transmission is complicated by the fact that messages need to be extracted from modulated signals, which are subjected to various distorting factors and interference in the radio channel. To solve this problem, a large number of methods have been developed and among them methods of digital signal reception based on neural network are beginning to develop [163].

However, systems with OFMn have disadvantages compared to AFMn: the need to transmit a pilot signal at the beginning of the communication session, less noise immunity, more complex hardware implementation. In addition, when a random jump distorts not only the current symbol, but also the next symbol after it, and the correction of a twofold error requires the use of a corrective code with a greater corrective power. Partially solves the problems of OFMn proposed in [164] method of differential space-time block coding, which has less computational complexity and allows to abandon pilot signals, which increases the efficiency of radio spectrum utilization. However, this method can be applied only in channels with relatively rare phase jumps, in which the coherence time is longer than the duration of two code blocks.

The solution to the problem of application of AFMN can be given by consideration of the phenomenon of "reverse work" in terms of noise-resistant coding. From this point of view "reverse work" is a set of packet errors, the number of which can reach 100% of the length of the transmitted codeword. One of the most effective of the known codes for packet error correction is the Reed-Solomon code (RS-code), widely used in noise-resistant coding. The limit of corrective ability (n,k) of RS-code is defined by Singleton's boundary, according to which for correction of t errors the code should have not less than $n-k=2t$ check symbols, i.e. two check symbols per one error. At a large degree of redundancy ($n>>k$) the number of corrected errors t approaches 50% of the codeword length n. However, the holographic code described in [44,165], which belongs to the family of positional divisible codes, is significantly more efficient. From the theory of electrical communication we know the dependence of the throughput of a binary symmetric channel on the probability of error in the channel, according to which the throughput is maximal at zero and unit error probability, and drops to zero at an error probability equal to 0.5. This case is called channel collapse. Indeed, an error probability of 0.5 can be obtained without transmitting information over the communication channel. And when the error probability equal to 1, the throughput is the same as at 0 (channel without interference). This is explained by the fact that in a binary channel to eliminate errors in all bits it is enough to invert them to absolutely completely restore the transmitted signal. The task of correcting errors in all bits of a binary codeword is trivial in itself - it is enough to invert each bit. The problem in this case for known codes is a choice from two equally probable results of decoding - direct and inverse. The holographic position code in contrast to other codes gives the coinciding result of decoding, both for direct, and for the inverted data block.

10.1 Holographic code and packet errors in "backtracking" mode

The holographic coding method is based on mathematical modeling of a one-dimensional hologram created in virtual space by a wave from an object obtained by transforming a k-bit binary code of the input data block into a secondary block - a single position code with the number of digits $n=2^k$. This transformation lays down information redundancy with the

number of digits r=n-k. The secondary block (virtual optical object) has
(n-1) zeros and one unit at the position given by the input block. Thus, the
input data block is used as the address of the unit position in the sequence
of zeros of the unit position code of the secondary block. The procedure
of formation of the hologram transmitted over the communication channel
and restoration of the original object by the hologram in the receiver is
described in [44].

The stability of the holographic code to errors is explained by the
property of hologram divisibility, according to which even at loss or
distortion of the most part of the hologram it is possible to restore the full
image of the object. The study of corrective ability of the holographic code
is carried out by modeling in MATLAB environment the process of
distortion of the hologram H by packet errors.

For a 5-bit (k=5) input data block A, the secondary block in the unit
position code has n=32 digits. For the value A recorded in decimal code
A_{10} =20, recording in binary code A_2 =10100, in unit position code

$$A \ \text{xml-ph-0000@deepl.internal}$$

$$=00000000000100000000000000000000. \qquad (1)$$

The digital one-dimensional hologram H obtained by the method
described in [44] is as follows:

$$H=01001110000000001110010010110101.$$

It is this block of data that is transmitted through the communication
channel. The result of the object recovery from the received hologram
(array Y) is shown in Figure 12.1 (the maximum is at the same position as
the unit in the position code (1) for A_1). The number of the Y array
position where its peak value is located, counted as in the numbering of
digits in the binary code from right to left, corresponds to the transmitted
value of the input block A=20.

The mode of "reverse work" (appearance of packet errors) leads to hologram inversion. Figure 12.2 shows the result of hologram decoding in the "reverse operation" mode with the number of errors equal to 100%. The peak value is in the same position. Therefore, to restore the value of the transmitted data block it is necessary only to determine the position of the global extremum irrespective of its sign.

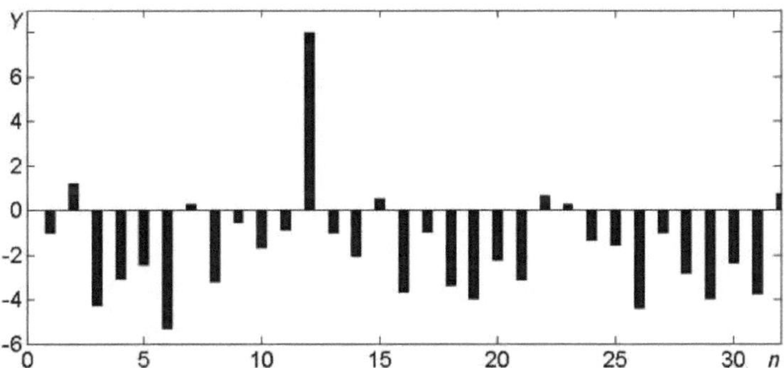

Fig. 12.1. The object reconstructed from the hologram

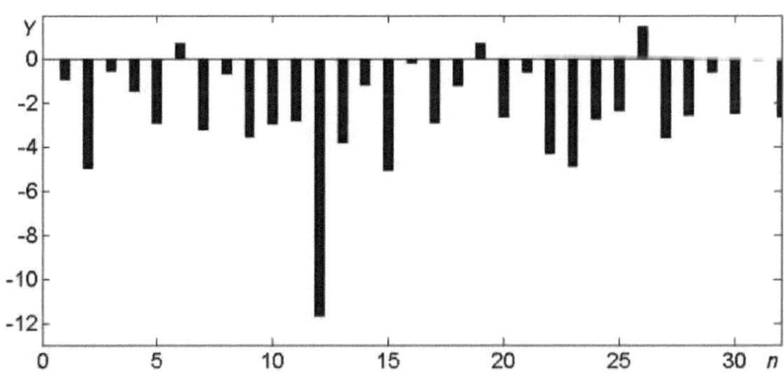

Fig. 12.2. Reconstructed array with the number of errors 32 (100%)

The above examples demonstrate the prospects of integrating the method of positional holographic noise-tolerant coding, which has the ability to eliminate any number of packet errors in the "reverse operation" mode, with the method of absolute phase manipulation.

Of great importance is the ratio of the coherence time (the duration of the interval between phase jumps that put the demodulator in the "reverse operation" mode and back) to the transmission time of one data block. If the coherence time is greater than the transmission time of a data block, then all blocks received during a stable phase of any sign are error-free decoded by a simple hard decoder. A data block during the reception of which a phase jump has occurred requires more complex, soft decoding.

10.2 Decoder development and simulation results

Thus, in the "reverse operation" mode the holographic decoder provides reliable operation of the demodulator. More difficult case is the moment of mode change - the moment of phase jump, when even relative phase manipulation gives two faulty bits. To solve this problem it is necessary to divide the received block into two parts and decode each half separately. The property of divisibility and the reserve of noise immunity of the holographic code caused by it allows to decode correctly a block of data not only on its half, but also on any quarter. The most difficult case for decoding is when the phase jump occurs twice during reception of one data block and duration of "reverse work" is about half of block length, i.e. the error packet is equal to half of the block and is located with an arbitrary offset in the block. In such a channel the OFMn, for example, forms 4 errors during demodulation of a data block.

To provide error-free decoding at any error packet length (from 0 to 100%), at any its position in the received block, a soft decoder is developed, using a more complex algorithm for analyzing the shape of the reconstructed signal. The algorithm contains two sets of operations - search for a special point (a point, in the number of position of which the transmitted information is embedded) by direct signs and search by indirect signs. Direct signs are maximum modulo values of 10 decoded arrays (direct and inverse forms of the full data block, direct and inverse forms of four quarters of the received data block). Indirect signs are characteristic systematic arrangement of maxima around a special point when its eigenvalue is minimal. When analyzing the received arrays, it is found that switching to the "inverse operation" mode leads to the replacement of the maximum at the special point by a group of two, four or six maxima arranged symmetrically around the special point. This task is suitable for a neural network, but it can also be accomplished with much smaller computational resources, including hard logic devices. Figure 12.3 shows examples of array fragments containing special points.

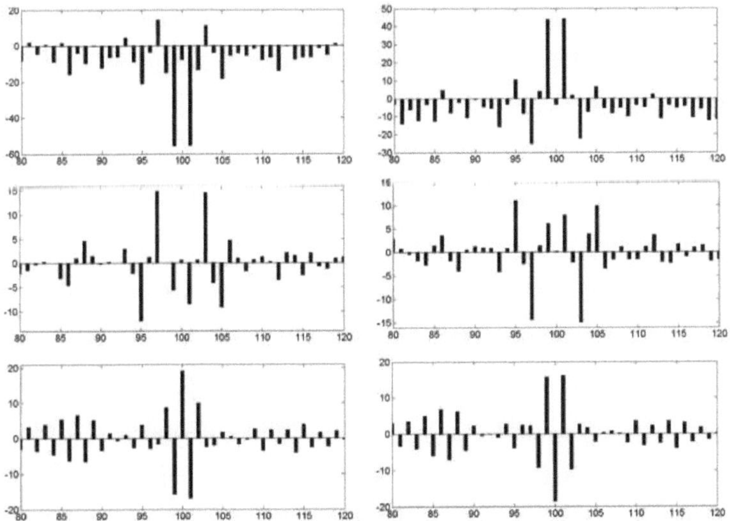

Fig. 12.3. Arrays obtained by decoding three parts of the received data block in direct and inverse codes. Coordinate of the special point A=100.

The soft holographic decoding algorithm includes the following operations:

1. Decoding the received array in direct code
2. Decoding an inverted array
3. Decoding the first quarter of direct and inverted arrays
4. Decoding the second quarter of direct inverted arrays
5. Decoding the third quarter of direct and inverted arrays
6. Decoding the fourth quarter of direct and inverted arrays
7. Search for positive maxima in each decoding result and determination of their coordinates
8. Search for negative maxima in each result and determine their coordinates
9. Search for second maxima (positive and negative) and determination of coordinates
10. Search for special points based on the selection of pairs of maxima located symmetrically at a distance of 2, 4 and 6 points. Candidate for decoding result is in the center between them
11. In the obtained array of special points, the determination of the coordinate of the point that occurs most often is the result of decoding.

The described algorithm is realized in MATLAB environment and the process of decoding of AFMn signal with duration n=32, 64, 128 and 256 bits at all possible sizes of error packets and all possible positions of the packet in the received signal was investigated. For all variants two phase jumps during reception of one data block were modeled. It is established, that the offered soft decoder in all these cases guarantees error-free formation of an output code.

The minimum length of the transmitted data block when using AFMn with holographic coding (AFMnGC) is 32 bits, this block is decoded in 5 information bits. When using OFMn with the same block length (32 bits) at the same intensity of phase failures (two jumps per block) 4 bits are incorrectly demodeled, i.e. 12.5% of errors occur. When using AFMnGC, errors start to appear only at three or more phase jumps per data block.

The proposed method of integration of absolute phase modulation with holographic coding allows to use the main advantage of AFMn - higher noise immunity, as well as simpler realization of modulation/demodulation. It eliminates the need for synchronization and the use of pilot signals. AFMnGK provides error-free decoding of signals carrying 5 information bits, and using for this 32 periods of carrier frequency (data rate density of 0.16 bits / Hz) in the coherence time, equal to 16 periods of the carrier.

List of references

1.   Optical holography / R. Collier, K. Burkhart, L. Lin. - M.: Mir, 1973.

2.   Chapurskiy V.V.. Obtaining of radio holographic images of objects on the basis of sparse antenna arrays of MIMO type with single-frequency and multi-frequency radiation / V.V. Chapurskiy. Chapurskiy // Bulletin of Bauman Moscow State Technical University. Ser. "Instrumentation". - 2011. №4.

3.   Semenchik, V.G. Radio holographic system of multi-frequency images formation (in Russian) / V.G. Semenchik, V.A. Pakhomov // Electronics. - 2004. № 1.

4.   Pahomov V. Reconstructing reflecting object images using born approximation / V. Pahomov. Pahomov, V. Semenchik, S. Kurilo // Proc. of 35th European Microwave Conference, - 2005. CNIT la Defense, Paris, France. P. 1375-1378.

5.   Shoidin, S.A. Synthesis of holograms at the receiving end of the communication channel with the holographic object / Shoidin, S.A. // Computer Optics, 2020, Vol. 44, No. 4 (DOI: 10.18287/2412-6179-CO-694).

6.   Timofeev, A.L. Using holographic coding to improve noise immunity of communication channels / Timofeev, A.L. // ITportal. - 2018. T.   18.   №2.   (http://itportal.ru/science/tech/ispolzovanie-golograficheskogo-kodi/).

7.   Golunov, V.A. Justification of the possibility of obtaining radio images of objects by the method of one-dimensional holograms / V.A. Golunov, V.A. Korotkov, K.V. Korotkov // Radio Engineering and Electronics, 2019, Vol. 64, No. 1 (DOI: 10.1134/S0033849419010066).

8.   Timofeev A.L.. Holographic method of error-correcting coding / A.L.. Timofeev, A.Kh. Sultanov // Proc. SPIE 11146. Optical Technologies for Telecommunications 2018. 111461A. N.Y.: SPIE, 2019. (https://doi.org/10.1117/12.2526922).

9.   Fundamentals of coding theory / Kudryashov B. D. - SPb.: BHV-Peterburg, 2016.

10.   Digital communication. Theoretical foundations and practical application. 2nd edition, revised / Sklyar B. - M.: Williams, 2007.

11.   Theory of error-correcting codes / McWilliams F. J., Sloan N. J. A. - M.: Svyaz, 1979.

12.   Timofeev A.L.. Holographic method for storage of digital information / A. L. L. Timofeev, A.Kh. Sultanov, P.E. Filatov // Proc. SPIE 11516, Optical Technologies for Telecommunications 2019, 1151604. N.Y.: SPIE, 2020. (doi:10.1117/12.2566329).

13.   Ciaramella E., Arimoto Y., Contestabile G., Presi M., D'Errico A., Guarino A., Matsumoto M. 1.28-Tb/s (32x40 Gb/s) free-space optical WDM transmission system // IEEE Photonics Technology Letters. 2009. V. 21. № 16. P. 1121-1123.

14.   Sahu M., Kiran K. V., Das S. K. FSO Link performance analysis with different modulation techniques under atmospheric turbulence // Proc. Second International Conference on Electronics, Communication and Aerospace Technology. Second International Conference on Electronics, Communication and Aerospace Technology. IEEE. 2018. P. 619-623.   DOI: 10.1109/ICECA.2018.8474849.

15.   Khalighi M. A., Uysal V. Survey on free space optical communication: a communication theory perspective // IEEE Communications Surveys & Tutorials. 2014. V. 16 № 8. P. 2231-2258. DOI:10.1109/COMST.2014.2329501.

16. Ghassemlooy Z., Popoola W.O.. Mobile and Wireless Communications Network Layer and Circuit Level Design, chapter Terrestrial Free-Space Optical Communications // InTech. 2010. P. 355-392.

17. Xu F., Khalighi M.A., Causs´e P., Bourennane S. Channel coding and time-diversity for optical wireless links // Optics Express. 2009. V. 17 № 2. P. 872-887.

18. Wilson S.G., Brandt-Pearce M., Cao Q.L., Baedke M. Optical repetition MIMO transmission with multipulse PPM // IEEE Journal on Selected Areas in Communications. 2005. V. 23. № 9. P. 1901-1910.

19. Gagliardi R.M., Karp S. Optical Communications. John Wiley & Sons, 2nd edition, 1995.

20. Xu F., Khalighi M.A., Bourennane S. Coded PPM and multipulse PPM and iterative detection for Free-Space optical links // IEEE/OSA Journal of Optical Communications and Networking. 2009. V. 1. № 5. P. 404-415.

21. Hemmati H. Deep Space Optical Communications. Wiley-Interscience, 2006.

22. Parfenov, V.I.; Golovanov, D.Yu. Interference immunity of the algorithms for receiving signals with multipulse position-pulse modulation (in Russian) // Computer Optics. 2018. T. 42 № 1. C. 167-174. DOI: 10.18287/2412-6179-2018-42-1-167-174.

23. Peppas K.P., Boucouvalas A.C., Ghassemloy Z. Performance of underwater optical wireless communication with multi-pulse pulse-position modulation receivers and spatial diversity // IET Optoelectronics. 2017. V. 11. № 5. P. 180-185. DOI: 10.1049/iet-opt.2016.0130.

24. Moision B., Hamkins J. Multipulse PPM on discrete memoryless channels // IPN Progress Report. Jet Propulsion Laboratory. 2005. V. 42. №. 160.

25. Hassan M., Shapsough S., Landolsi T., Elrefaie A.F.. Error performance study of MPPM optical communication systems with finite extinction ratios // Proceedings of International Conference on Industrial Informatics and Computer Systems. 2016. P. 1-4. DOI: 10.1109/ICCSII.2016.7462445.

26. Simon M. K., Vilnrotter V. A. Performance analysis and tradeoffs for dual-pulse PPM on optical communication channels with direct detection // IEEE Transactions on Communications. 2004. V.52. № 11. P. 1969-1979.

27. Fan Y., Green R. J. Comparison of pulse position modulation and pulse width modulation for application in optical communications // Optical Engineering. 2007. V. 46. № 6.

28. Krasnov, R.P. System of atmospheric optical communication of OFDM-type on the basis of LDPC code with displacement in a turbulent channel // Bulletin of Voronezh State Technical University. 2018. T. 14. № 4. C. 71-76.

29. Bukola D. Ajewole, Kehinde O. Odeyemi, Pius A. Owolawi, and Viranjay M. Srivastava. Performance of OFDM-FSO Communication System with Different Modulation Schemes over Gamma-Gamma Turbulence Channel // Journal of Communications. 2019. V. 14. № 6. P. 490-497.

30. Ruhin Chowdhury, A. K. M. Sharoar Jahan Choyon. Design of Novel Hybrid CPDM-CO-OFDM FSO Communication System and its Performance Analysis under Diverse Weather Conditions // Journal of Optical Communications. 2021.   DOI: 10.1515/joc-2021-0113.

31.   Tang Q., Li K., Liu X., Kong L. Performance Analysis of Spectral Efficiently Adaptive Modulation DFT-Spread Polar Coordinate-Based OFDM in Hybrid Fiber-Visible Laser Light Communication System // 2020 IEEE 20th International Conference on Communication Technology. P. 594-598.

32.   Proakis J. G., Salehi M. Digital Communications. McGraw-Hill, New York, 5th edition, 2007.

33.   Khalighi M. A., Schwartz N., Aitamer N., Bourennane S. Fading reduction by aperture averaging and spatial diversity in optical wireless systems // IEEE/OSA Journal of Optical Communications and Networking. 2009. V. 1. № 6. P. 580-593.

34.   Anguita J. A., Djordjevic I. B., Neifeld M. A., Vasic B. V. Shannon capacities and error-correction codes for optical atmospheric turbulent channels // Journal of Optical Networking. 2005. V. 4. № 9. P. 586-601.

35.   Cvijetic N., Wilson S. G., Zarubica R. Performance evaluation of a novel converged architecture for digital-video transmission over optical wireless channels // IEEE/OSA Journal of Lightwave Technology. 2007. V. 25. № 11. P. 3366-3373.

36.   Djordjevic I. B., Denic S., Anguita J., Vasic B., Neifeld M. A. LDPC-coded MIMO optical communication over the atmospheric turbulence channel // IEEE/OSA Journal of Lightwave Technology. 2008. V. 26. № 5. P. 478-487.

37.   Djordjevic I. B., Vasic B., Neifeld M. A. LDPC coded OFDM over the atmospheric turbulence channel // Optics Express. 2007. V. 15 № 10. P. 6336-6350.

38.   Uysal M., Li J., Yu M. Error rate performance analysis of coded Free-Space Optical links over Gamma-Gamma atmospheric turbulence

channels // IEEE Transactions on Wireless Communications. 2006. V. 5. № 6. P. 1226-1233.

39. Chan V. W. S. Free-space optical communications // IEEE/OSA Journal of Lightwave Technology. 2006. V. 24. № 12. P. 4750-4762.

40. Shapiro J. H., Puryear A. L. Reciprocity-enhanced optical communication through atmospheric turbulence, Part I: Reciprocity proofs and far-field power transfer optimization // IEEE/OSA Journal of Optical Communications and Networking. 2012. V. 4. № 12. P. 947-954.

41. Puryeara A. L., Shapirob J. H., Parenti R. R. Reciprocityenhanced optical communication through atmospheric turbulence-Part II: Communication architectures and performance // IEEE/OSA Journal of Optical Communications and Networking. 2013. V. 5. № 8. P. 888-900.

42. Muhammad S. S., Javornik T., Jelovcan I., Leitgeb E., Ghassemlooy Z. Comparison of hard-decision and soft-decision channel coded M-ary PPM performance over free space optical links // European Transactions on Telecommunications. 2008. V. 20. № 8. P. 746-757.

43. Timofeev A.L., Sultanov A.H. Positional divisible codes with error correction // In Proc. Problems of Telecommunication Techniques and Technologies PT&T-2019. Kazan, 2019. C. 132-135.

44. Timofeev A.L., Sultanov A.H. Construction of noise-resistant code on the basis of holographic representation of arbitrary digital information // Computer Optics. 2020, T. 44, № 6. C. 978-984. DOI: 10.18287/2412-6179-CO-739.

45. Kaushal H., Jain V., Kar S. Free-Space Optical Channel Models // Free Space Optical Communication. P. 41-89. DOI:10.1007/978-81-322-3691-7_2.

46. Huihua F.U., Wang P., Liu T., Cao T., Guo L., Qin J. Performance analysis of a PPM-FSO communication system with an avalanche

photodiode receiver over atmospheric turbulence channels with aperture averaging // Applied Optics. 2017. V. 56, № 23. P. 6432-6439. DOI: 10.1364/AO.56.006432.

47. Gappmair W., Hranilovic S., Leitgeb E. Performance of PPM on Terrestrial FSO Links with Turbulence and Pointing Errors // IEEE Communications letters. 2010. V. 14, №. 5.

48. Zubov, V.A. Analysis of time-varying optical signals and transfer functions using spectral modulation / Zubov V.A., Merkin A.A. // Quantum Electronics, 1999, Vol. 29, No. 2

49. Mazurenko Yu.T. Spectral holography / Mazurenko Yu.T. // Optical Journal, 1994, Vol. 61, No. 1.

50. Kalinin, V.I. Information transmission based on noise signals with spectral modulation / Kalinin V.I., Chapursky V.V. // Radio Engineering and Electronics, 2015, Vol. 60, No. 10.

51. Zlobin, V.K. Spectral methods of image processing / V.K. Zlobin, B.V. Kostrov, A.S. Asaev, E.R. Muratov // Vestnik of RGRTU. Vop. 21. Ryazan, 2007.

52. Parfenov, V.I. Interference immunity of the algorithms for receiving signals with multi-pulse position-pulse modulation / Parfenov, V.I.; Golovanov, D.Yu. // Computer Optics. - 2018. - T. 42, № 1. - C. 167-174. - DOI: 10.18287/2412-6179-2018-42-1-167-174.

53. Dufaux F., Xin, Y., Pesquet-Popescu B., Schelkens P.. Compression of digital holographic data: An overview. Proc. SPIE 9599, 2015. doi:/10.1117/12.2190997.

54. Hajihashemi V., Najafabadi H.E., Gharahbagh A.A., Leung H., Yousefan M.Y., Tavares J.M. A novel high-efficiency holography image compression method based on HEVC, Wavelet, and nearest-neighbor

interpolation. Multimedia Tools and Applications 2021; 80: 31953-31966. doi:10.1007/s11042-021-11232-0.

55. Peixeiro J.P., Brites C., Ascenso J.A., Pereira F. Holographic data coding: benchmarking and extending HEVC with adapted transforms. IEEE Transactions on Multimedia. 2018; 20(2): 282-297. doi:10.1109/TMM.2017.2742701.

56. Cheremkhin P.A., Kurbatova E.A. Wavelet compression of off-axis digital holograms using real/imaginary and amplitude/phase parts. Scientific Reports. 2019; 9. doi:/10.1038/s41598-019-44119-0.

57. Narayanan R.M., Chuang J. Electron. Lett. 2007; 43(22): 1211.

58. Liang Shi, Beichen Li, Changil Kim, Petr Kellnhofer & Wojciech Matusik. Towards real-time photorealistic 3D holography with deep neural networks. Nature. 2021; 591(11): 234-253. doi:10.1038/s41586-020-03152-0.

59. Scherbakov M.A., Panov A.P. Nonlinear filtering with adaptation to local image properties. Computational Optics. 2014; 38(4): 818-824.

60. Eisenberg I.N. Some algorithms of image processing and their realization on neural networks. Computer Optics. 1997; 17: 134-142.

61. Selomon D. Compression of data, images and sound. Moscow: Technosphere, 2006.

62. Grishentsev A.Yu. Efficient image compression on the basis of differential analysis. Journal of radio electronics. 2012; 11.

63. Coifman R.R., Wickerhauser M.V. Entropy-based algorithms for best basis selection. IEEE Trans. Inform. Theory, Special Issue on Wavelet Transforms and Multires. Signal Anal 1992; 38: 713-718.

64. Umnyashkin S.V., Gizatullin R.R. Image compression on the basis of block decomposition in the domain of packet wavelet transform. Digital Signal Processing. 2014; 1.

65. Saupe D., Hamzaoui R., Hartenstein H. Fractal image compression - An introductory overview, Fractal Models for Image Synthesis, Compression, and Analysis, D. Saupe, J. Hart (eds.), ACM SIGGRAPH'96 Course Notes.

66. Jiao, S. et al. Compression of phase-only holograms with JPEG standard and deep learning. Appl. Sci. 2018; 8: 1258.

67. Glumov N.I. Complex approach at a choice of algorithms of compression and noise-resistant coding for transmission of digital images on communication channels. Computer optics. 2004; 26: 106-109.

68. Gashnikov M.V., Glumov N.I. Increase of compression ratio and visual quality at hierarchical image compression due to preliminary filtering. Computer Optics 2005; 28: 108-111.

69. Demin V.V., Kozlova A.V. Methods of coding-decoding of digital holograms of particles. Izvestia vuzov. Physica. 2013; 56(10): 368-371.

70. Donoho D. L. Compressed Sensing. IEEE Transactions on Information Theory 2006; 52(4): 1289-1306.

71. Eldar Y. C. Sampling theory: Beyond bandlimited systems. Cambridge University Press 2015.

72. Fadeev D.K., Rashich A.V. Optimal input power backoff of a nonlinear power amplifier for SEFDM-system. Proceedings of the NEW2AN 2015 and 8th Conference, 2015, 669-678.

73. Tom A. Suppressing alignment: Joint PAPR and out-of-band power leakage reduction for OFDM-based systems. Department of Electrical Engineering, University of South Florida, Tampa, FL, 2016.

74. Isam S., Darwazeh I. Peak to average power ratio reduction in spectrally efficient FDM systems. Proceedings of the 18th International Conference on Telecommunications, 2011.

75.   Timofeev A.L., Sultanov A.Kh. Influence of noise and sampling frequency on the error of discrete image representation. Information and control systems. 2021; 5: 31-38. doi:10.31799/1684-8853-2021-5-33-39.

76.   Lidia Galdino L., Edwards A., Yi W., Sillekens E., Wakayama Y., Gerard T., Pelouch W.,S., Barnes S., Tsuritani T., Killey R., I., Lavery D., Bayvel P. Optical Fibre Capacity Optimization via Continuous Bandwidth Amplification and Geometric Shaping // IEEE Photonics Technology Letters. 2020. V. 32, no. 17, pp. 1021-1024. doi: 10.1109/LPT.2020.3007591

77.   Grigorieva, E.E.; Semenov, A.T. Waveguide transmission of images in coherent light (review) // Quantum Electronics. 1978. T. 5. № 9. C. 1877-1895.

78.   Richardson D.J., Fini J.M., Nelson L.E.. Space-division multiplexing in optical fibers // Nature Photonics. 2013. V. 7. № 5. P. 354-362. doi: 10.1038/nphoton.2013.94

79.   Wright L.G., Christodoulides D.N., Wise F.W.. Controllable spatiotemporal nonlinear effects in multimode fibers // Nature Photonics. 2015. V. 9. P. 306-310. doi: 10.1038/nphoton.2015.61

80.   Cizmar T., Dholakia K.. Exploiting multimode waveguides for pure fiber-based imaging // Nature Communication. 2012. V. 3. P.1027.

81.   Choi Y., Yoon C., Kim M., Yang T. D., Fang-Yen C., Dasari R. R., Lee K. J., Choi W. Scanner-free and wide-field endoscopic imaging by using a single multimode optical fiber // Physical Review Letters. 2012. V. 109. № 20.

82.   Turtaev S., Leite I.T., Altwegg-Boussac T., Pakan J.M., Rochefort N.L., Cizmar T. High-fidelity multimode fiber-based endoscopy for deep brain in vivo imaging // Light: Science and Applications. 2018. V. 7. № 1.

83.　Resisi S., Popoff S. M., Bromberg Y. Image transmission through a dynamically perturbed multimode fiber by deep learning // Laser & Photonics Reviews. 2021. № 10. doi:10.48550/arXiv.2011.05144

84.　Lucesoli A., Rozzi T. Image transmission by multimode optical fiber for microendoscopy // Proc. of SPIE-OSA Biomedical Optics. 2007. SPIE V. 6631. 663117.

85.　Caramazza P., Moran O., Murray-Smith R., Faccio D.. Transmission of natural scene images through a multimode fiber // Nature Communications. 2019.　V. 10. № 2029. doi.org/10.1038/s41467-019-10057-8

86.　Fertman A., Yelin D.. Image transmission through an optical fiber using real-time modal phase restoration // Journal of the Optical Society of America B. 2013. V. 30. № 1. P. 149-157. doi.org/10.1364/JOSAB.30.000149

87.　Bailey D., Wright E. Practical Fiber Optics: Elsevier. IDC Technologies. 2003. 245 p.

88.　Ho K., Kahn J. Mode Coupling and its Impact on Spatially Multiplexed Systems. Optical Fiber Telecommunications VIB: Systems and Networks: Sixth Edition. 2013. doi.org/10.1016/B978-0-12-396960-6.00011-0.

89.　Barankov R., Mertz J. High-throughput imaging of self-luminous objects through a single optical fiber // Nature Communications. 2014. V. 5. no. 5581 doi.org/10.1038/ncomms6581.

90.　Feshchenko, V.S.; Rogozhnikova, O.A. Optical imaging system with a waveguide (in Russian) // Optics and Spectroscopy. 2004. T. 97. № 3. C. 498-501.

91.　Bakharev, M.A.; Kotlyar, V.V.; Pavel'ev, V.S.; Soifer, V.A.; Khonina, S.N. Effective excitation of mode packets of an ideal gradient waveguide with specified phase velocities // Computer Optics. 1997. № 17. С. 21-25.

92. Liu C., Deng L., Liu D., Su L. Modeling of a single multimode fiber imaging system // arXiv:1607.07905 [physics.optics]. doi.org/10.48550/arXiv.1607.07905.

93. Kakkava E., Rahmanib B., Borhania N., Tegina U., Loterieb D., Konstantinoub G., Moserb C., Psaltis D.. Imaging through multimode fibers using deep learning: The effects of intensity versus holographic recording of the speckle pattern // Optical Fiber Technology. 2019. V. 52. 101985. doi.org/10.1016/j.yofte.2019.101985.

94. Borhani N., Kakkava E., Moser C., Psaltis D. Learning to see through multimode fibers // Optica. 2019 V. 5. № 8, P. 960-966. doi.org/10.1364/OPTICA.5.000960.

95. Fan P., Zhao T., Su L. Deep learning the high variability and randomness inside multimode fibers // arXiv:1807.09351 [physics.optics]. doi.org/10.48550/arXiv.1807.09351.

96. Rahmani B., Loterie D., Konstantinou G., Psaltis D., Moser C. Multimode optical fiber transmission with a deep learning network // Nature. Light Appl. 2018. V. 7. № 69. doi.org/10.1038/s41377-018-0074-1.

97. Takagi R., Horisaki R. & Tanida J. Object recognition through a multi-mode fiber // Opt Rev. 2017. № 24. P. 117-120. doi.org/10.1007/s10043-017-0303-5.

98. Pauwels, J., Van der Sande, G. & Verschaffelt, G. Space division multiplexing in standard multi-mode optical fibers based on speckle pattern classification. Sci Rep 9, 17597 (2019). doi.org/10.1038/s41598-019-53530-6.

99. Lei Y., Li J., Fan Y., Yu D., Fu S., Yin F., Dai Y., Xu K. Space-division-multiplexed transmission of 3x3 multiple-input multiple-output wireless signals over conventional graded-index multimode fiber. Opt. Express. 2016. № 24, P. 28372-28382. doi.org/10.1364/OE.24.028372.

100. Mohapatra H., Hosain S. Intermodal dispersion free few-mode (quadruple mode) fiber: A theoretical modeling. Opt. Commun. 2013. № 30. P. 267-270. doi.org/10.1016/j.optcom.2013.05.018.

101. Kubota H., Morioka T. Few-mode optical fiber for mode-division multiplexing. Opt. Fiber Technol. 17, 490-494 (2011). doi.org/10.1016/j.yofte.2011.06.011.

102. Timofeev, A.L.; Sultanov, A.Kh.; Meshkov, I.K.; Gizatulin, A.R. Range increase of the atmospheric optical communication lines by means of position coding (in Russian) // Optical Journal. 2022. T. 89. № 9. C. 75-85. doi:10.17586/1023-5086-2022-89-09-75-85.

103. Leonardo R. D., Bianchi S. Hologram transmission through multi-mode optical fibers. Opt. Express. 2011. № 19. P. 247-254. doi:10.1364/OE.19.000247

104. Paurisse M., Hanna M., Droun F., Georges P., Bellanger C., Brignon A., Huignard J. P. Phase and amplitude control of a multimode fiber beam by use of digital holography. Opt. Express. 2009. № 17. P. 13000–13008. doi:10.1364/OE.17.013000.

105. Leonardo R.D., Ianni F., Ruocco G. Computer generation of optimal holograms for optical trap arrays. 2007. Opt. Express. № 15. P. 1913–1922. doi:10.1364/OE.15.001913.

106. Grier D.G. A revolution in optical manipulation. Nature. 2003. № 424. P. 810-816. doi:10.1038/nature01935.

107. Spalding G.C., Courtial J., Leonardo R.D. Holographic optical tweezers. Structured Light Its Applications. Academic Press. 2008. P. 139–168. doi:10.1016/B978-0-12-374027-4.00006-2.

108. Reicherter M., Haist T., Wagemann E.U., Tiziani H.J.. Optical particle trapping with computer-generated holograms written on a liquid-crystal display. Opt. Lett. 1999. № 24, P. 608-610. doi:10.1364/OL.24.000608.

109. Maximov, I. A.; Kochura, S. G.; Avdyushkin, S. A. Basic provisions of the methodology for ensuring the resistance of onboard spacecraft hardware to the radiation effects of space // Siberian Aerospace Journal. A. Basic provisions of the methodology for ensuring the resistance of spacecraft onboard equipment to the radiation effects of space // Siberian Aerospace Journal. 2023. T. 24, № 1. C. 116-125. Doi: 10.31772/2712-8970-2023-24-1-116-125.

110. Collier R. J., Burckhardt C. B., Lin L. H. Optical Holography // Murray Hill, New Jersey. 1971.

111. Bruckstein A.M., Holt R.J., Netravali A.N.. Holographic image representations: the subsampling method // IEEE Int. Conference on Image Processing. Santa Barbara. California, USA. Vol. 1, 177-180, 1997.

112. Bruckstein A.M., Holt R.J. Netravali A.N. Holographic representation of images // IEEE Transactions on Image Processing. 1998. No. 7. P. 1583-1587.

113. Bruckstein A.M., Holt R.J., Netravali A.N.. On Holographic Transform Compression of Images // Proceedings 15th International Conference on Pattern Recognition ICPR-2000. John Wiley & Sons Inc. P. 244-252, 2001. DOI: 10.1109/ICPR.2000.903528.

114. Dovgard R. Holographic image representation with reduced aliasing and noise effects // Image Processing. IEEE Transactions. 2004. No. 13(7), P. 867-872.

115. Kolesov, V.V.; Zalogin, N.N.; Vorontsov, G.M. Pseudoholographic coding method (in Russian) // Radio engineering and electronics. 2002. T. 2, № 5. C. 583.

116. Barinova, D.A. Development and research of algorithms for processing digital images represented in pseudoholographic codes // Computer Optics. 2005. VOL. 27. P. 149-154.

117. Clark G.C. Jr., Cain J.B.. Error-Correction Coding for Digital Communications // Plenum Press, New York. Second printing, 1982.

118. Timofeev A.L., Sultanov A.H. Application of noise-resistant position divisible codes // In Proceedings of the XXII International Scientific and Technical Conference "Problems of Telecommunications Engineering and Technology-2020". Samara: PGUTI, 2020. C. 12-15.

119. Gallagher R. Information Theory and Reliable Communication // New York: Wiley, 1968.

120. Anderson J.B., Mohan S. Source And Channel Coding An Algorithmic Approach // Springer Science+Business Media. New York. 1991.

121. Sklar B. Digital Communications: Fundamentals and Applications. Second Edition // Prentice Hall P T R Upper Saddle River. New Jersey. 2001.

122. Mac Williams F.J., Sloane N.J.A. The Theory of Error-Correction Codes // Bell Laboratories. Murray Hill. NJ 07974. U.S.A. 1977.

123. Pudlovskiy, V.B. Selection of GLONASS satellites for reducing the error in determining the planned coordinates // Rocket and Space Instrumentation and Information Systems. 2019. Vol. 6. No. 3. C. 15-22.

124. Aleshin, B.S.; Antonov, D.A.; Veremeenko, K.K.; Zharkov, M.V.; Zimin, R.Yu.; Kuznetsov, I.M.; Pronkin, A.N. Small-size integrated navigation and landing complex // Proceedings of MAI. 2012. № 54. URL: https://www.mai.ru/science/trudy/published.php?ID=29692

125. Valaitite A.A., Nikitin D.P., Sadovskaya E.V. Investigation of the influence of multipath error on the accuracy of GNSS (Global Navigation Satellite Systems) signal parameters determination using a navigation field simulator // Proceedings of MAI. 2014. № 77. URL: http://www.mai.ru/science/trudy/published.php?ID=53172

126. Г. N. Maltsev, A. N. Sakulin, E. A. Sakulin. Potential accuracy of mobile measuring points georeferencing using GLONASS satellite navigation system signals // Voprosy Radioelectroniki, ser. Technika Television. 2015. № 2. C. 57-64.

127. Ryabov, I.V.; Romanov, I.S. Determination of factors affecting the accuracy of positioning using global navigation satellite systems GPS and GLONASS // DSPA: Issues of digital signal processing application. 2018. T. 8. № 2. C. 167-170.

128. H. C. Tsyrempilova, T. E. Khavronina. Accuracy of measurement of navigation parameters in the navigation equipment of the glonass satellite radionavigation system consumer equipped with an antenna array // Actual problems of aviation and astronautics. 2015. T. 1.   C. 80-82.

129. П. V. Sharshavin, A. S. Kondratyev, A. V. Grebennikov. Application of digital registration to improve the accuracy characteristics of pseudodistance measurement by GLONASS/GPS satellite radionavigation systems signals // Bulletin of the Siberian State Aerospace University named after academician M. F. Reshctncv. 2012. Vop. 1 (41). C. 109-111.

130. Mishin, A.Yu.; Frolova, O.A.; Isaev, Yu.K.; Egorov, A.V. Integrated navigation system of an aircraft // Proceedings of MAI. 2010. № 38.

131. Ivanov, V.F.; Koshkarov, A.S. Interference immunity enhancement of the GLONASS consumer navigation equipment due to complexing with inertial navigation sensors // Proceedings of MAI. 2017. Issue No. 93. C. 23- 39.

132. Tkachev, A.B. New ways to improve noise immunity of signals of global navigation satellite systems // MAI Bulletin. 2011.   T. 18. № 5. C. 72-77.

133. B. V. Dvorkin, R. V. Bakitko, V. V. Kurshin, A. A. Povalyaev. Russian navigation and information satellite system // Rocket and Space Instrumentation and Information Systems. 2018. T. 5, issue 3. C. 3-16. DOI 10.30894/issn2409-0239.2018.5.3.3.16.

134. GLONASS. Interface control document (version 5.1). - Moscow: Russian Research Institute of Space Instrumentation, 2008. 60 c.

135. Soloviev Yu.A. Satellite navigation systems. M.: ECO-TRENDS, 2000. 268 c.

136. Gorgadze, S. F., Ermakova, A. V., Efficiency of multiple access options for 5G and 6G cellular networks. H&ES Reserch, 2022. Vol. 14. no 2. P. 19-26. doi: 10.36724/2409541920221421926.

137. Fertman A., Yelin D. Image transmission through an optical fiber using real-time modal phase restoration // J. Opt. Opt. Soc. Am. 2013; 30(1): 149-157.

138. Lucesoli A., Rozzi T. Image transmission by multimode optical fiber for microendoscopy // Proc. of SPIE-OSA Biomedical Optics, SPIE; 2016; Vol. 6631, 663117.

139. Rahmani B., Oguz I., Tegin U., Hsieh J., Psaltis D., Moser C. Learning to image and compute with multimode optical fibers // Nanophotonics 2022; 11(6): 1071-1082.

140. Shoidin S.A., Pazoev A.L., Smyk A.F., Shurygin A.V. Synthesized at the receiving end of the communication channel 3D-object holograms in Dot Matrix technology // Computer Optics. - 2022. - Т. 46, № 2. - C. 204-213. - DOI: 10.18287/2412-6179-CO-1037.

141. Panuski, C.L., Christen, I., Minkov, M. Brabec C.J., Trajtenberg-Mills S., Griffiths A.D., McKendry J.D., Leake G.L., Coleman D.J., Tran C., Louis J.S., Mucci J., Horvath C., Westwood-Bachman J.N., Preble S.F., Dawson M.D., Strain M.J., Fanto M.L., Englund D.R.. A full degree-of-freedom spatiotemporal light modulator // Nature Photonics, 2022, no. 16, pp. 834-842. doi.org/10.1038/s41566-022-01086-9.

142. Timofeev, A.L.; Sultanov, A.Kh.; Meshkov, I.K.; Gizatulin, A.R. Use of the holographic methods of image transmission over the multimode optical fiber to increase the capacity of fiber-optic communication lines // Optical Journal. 2023. Т. 90. № 10. C. 13-23. http://doi.org/10.17586/1023-5086-2023-90-10-13-23.

143. Huang C., Hu S., Alexandropoulos G.C., Alessio Zappone A., Yuen C., Zhang R., Di Renzo M., Debbah M. Holographic MIMO Surfaces for 6G Wireless Networks: Opportunities, Challenges, and Trends // IEEE Wireless Communications 2020; 27(5). DOI: 10.1109/MWC.001.1900534.

144. Molotkov S.N. About superluminal group velocity and information transfer. Letters in ZhETF. 2010. Т. 91. № 12. C. 762-768.

145. Oshchepkov P. K., Merkulov A. P., Introscopy, M., 1967.

146. Irisova N. A., Timofeev Yu. P. Radio vision of ground objects in complex meteorological conditions, Moscow, 1969;

147. Friedman, S. A., Luminescence allows us to see the invisible, Nature, 1975, no. 1.

148. Dmitriev A.S., Petrosyan M.M., Ryzhov A.I. Experimental model of a multibeam device for observation in radio light // Letters in ZhtF, 2021, vol. 47, issue 12, pp. 38-41. DOI: 10.21883/PJTF.2021.12.51066.18762.

149. Dmitriev A.S., Efremova E.V., Gerasimov M Yu, Itskov V.V.. Radio illumination based on ultra-wideband dynamic chaos generators // Radio Engineering and Electronics, 2016, Vol. 61, No. 11, pp. 1073-1083. DOI: 10.7868/S0033849416110024.

150. Karanam C. R., Mostofi Y. 3D Through-Wall Imaging with Unmanned Aerial Vehicles Using WiFi // 16th ACM/IEEE International Conference on Information Processing in Sensor Networks (IPSN), 2017. P. 131-142.

151. Korany B., Karanam C. R., Mostofi Y. Adaptive Near-Field Imaging with Robotic Arrays // IEEE Sensor Array and Multichannel Signal Processing Workshop (SAM), 2018.

152. Ivashov C.B., Bugaev A.S. Use of noise generators in radiometric systems for detection of hidden objects // Radiotekhnika i elektronika, 2013, vol. 58, no. 9, pp. 935-942 DOI: 10.7868/S0033849413090052.

153. Thompson A. R., Moran J. M., Swenson G. W. Interferometry and Synthesis in Radio Astronomy // 3rd Edition, 2017. DOI:10.1007/978-3-319-44431-4.

154. B.A. Panchenko, D.V. Denisov. Diffraction characteristics of Luneberg lens for circular polarization field // Physics of Wave Processes and Radio Engineering Systems, 2013. vol. 16, № 4.

155. Kusaykin, D.V.; Denisov, D.V. Channel estimation in 5G MIMO-OFDM systems with multibeam lens antennas // Vestnik SibGUTI, 2021, No. 4.

156. Panchenko B.A., Ponomarev O.P., Denisov D.V. Fast calculation of the scattering characteristics of the Luneberg lens // Bulletin of VKO Concern Almaz-Antey, 2017, no. 2, pp. 21-25. DOI:10.38013/2542-0542-2017-2-21-25.

157. Denisov D. Highly Directional Luneberg Lensed Antennas. https://nag.ru/material/28514.

158. Basov N.G. On the Way to Optical Radio. Nauka i zhizn. 1961. NO. 7. https://www.nkj.ru/archive/articles/42954/.

159. Golovin O. Relative phase manipulation - a method of increasing the reliability of information transmission. URL: https://www.computer-museum.ru/connect/petrovic.htm (date of reference: 10.04.2024).

160. Volkov, A.A.; Morozov, M.S. Method of realization of absolute FMn at 180° . // Electrosvyaz. - 2017. - № 2. C. 72-74.

161. Petrovich N.T. New ways of realization of phase telegraphy. // Radiotekhnika. - 1957. - № 10.

162. Kulikov, G.V.; Do, C.T. Efficiency of the phase adaptive filtering algorithm for receiving signals with multi-position phase manipulation. // Journal of Radio Electronics. - 2020. - №. 4. https://doi.org/10.30898/1684-1719.2020.4.9.

163. Evstratko, V.V.; Konovalenko, A.I.; Mishurov, A.V.; Yukhmanov, A.D. Digital radio receiver based on a neural network. // Journal of Radioelectronics. - 2024. - №. 1. https://doi.org/10.30898/1684-1719.2024.1.5.

164. Tokar, M.S.; Ryabov, I.V. Method of differential space-time block coding for application in mobile radio communication systems using

MIMO technology. // Journal of Radio Electronics. - 2021. - № 6. https://doi.org/10.30898/1684-1719.2021.6.4.

165. Timofeev A.L., Sultanov A.H., Meshkov I.K., Gizatulin A.R. Increase of active utilization period of onboard electronic equipment of spacecrafts. // Siberian Aerospace Journal. - 2024. - T. 25. - № 1. C. 33-42. http://doi.org/10.31772/2712-8970-2024-25-1-33-42.

More
Books!

info@omniscriptum.com
www.omniscriptum.com
OMNIScriptum

Printed by Books on Demand GmbH, Norderstedt / Germany